This book provides an easily accessible introduction to quantum field theory via Feynman rules and calculations in particle physics. The aim is to make clear what the physical foundations of present day field theory are, to clarify the physical content of Feynman rules, and to outline their domain of applicability. The book begins with a brief review of some aspects of Einstein's theory of relativity that are of particular importance for field theory, before going on to consider the relativistic quantum mechanics of free particles, interacting fields, and particles with spin. The techniques learnt in these chapters are then demonstrated in examples that might be encountered in real accelerator physics. Further chapters contain discussions on renormalization, massive and massless vector fields and unitarity. A final chapter presents concluding arguments concerning quantum electrodynamics. The book includes valuable appendices that review some essential mathematics, including complex spaces, matrices, the CBH equation, traces and dimensional regularization. An appendix containing a comprehensive summary of the rules and conventions used is followed by an appendix specifying the full Langranian of the Standard Model and the corresponding Feynman rules. To make the book useful for a wide audience a final appendix provides a discussion on the metric used, and an easy-to-use dictionary connecting equations written with a different metric. Written as a textbook, many diagrams and examples are included.

This book will be used by beginning graduate students taking courses in particle physics or quantum field theory, as well as by researchers as a source and reference book on Feynman diagrams and rules.

Diagrammatica

The Path to Feynman Rules

MARTINUS VELTMAN

University of Michigan

CAMBRIDGE
UNIVERSITY PRESS

CAMBRIDGE UNIVERSITY PRESS
Cambridge, New York, Melbourne, Madrid, Cape Town, Singapore,
São Paulo, Delhi, Dubai, Tokyo, Mexico City

Cambridge University Press
The Edinburgh Building, Cambridge CB2 8RU, UK

Published in the United States of America by
Cambridge University Press, New York

www.cambridge.org
Information on this title: www.cambridge.org/9780521456920

First published 1994
Reprinted 1995

A catalogue record for this publication is available from the British Library

Library of Congress cataloguing in publication data available

ISBN 978-0-521-45692-0 Paperback

CAMBRIDGE LECTURE NOTES IN PHYSICS 4

General Editors: P. Goddard, J. Yeomans

Diagrammatica
The Path to Feynman Rules

CAMBRIDGE LECTURE NOTES IN PHYSICS 4

Contents

vii

Introduction

In recent years particle theory has been very successful. The theory agrees with the data wherever it could be tested, and while the theory has its weak spots, this numerical agreement is a solid fact. Physics is a quantitative science, and such agreement defines its validity.

It is a fact that the theory, or rather the successful part, is perturbation theory. Up to this day the methods for dealing with non-perturbative situations are less than perfect. No one, for example, can claim to understand fully the structure of the proton or the pion in terms of quarks. The masses and other properties of these particles have not really been understood in any detail. It must be added that there exists, strictly speaking, no sound starting point for dealing with non-perturbative situations.

Perturbation theory means Feynman diagrams. It appears therefore that anyone working in elementary particle physics, experimentalist or theorist, needs to know about these objects. Here there is a most curious situation: the resulting machinery is far better than the originating theory. There are formalisms that in the end produce the Feynman rules starting from the basic ideas of quantum mechanics. However, these formalisms have flaws and defects, and no derivation exists that can be called satisfactory. The more or less standard formalism, the operator formalism, uses objects that can be proven not to exist. The way that Feynman originally found his diagrams, by using path integrals, can hardly be called satisfactory either: on what argument rests the assumption that a path integral describes nature? What is the physical idea behind that formalism? Path integrals are objects very popular among mathematically oriented theorists, but just try to sell them to an experimentalist. However, to be more positive, given that one believes Feynman diagrams, path integrals may be considered a very valuable tool to understand properties of these diagrams. They are justified by the result, not by their definition. They are mathematical tools.

Well, things are as they are. In this book the object is to derive Feynman rules, but there is no good way to do that. The physicist may take a pragmatic attitude: as long as it works, so what. Indeed, that is a valid attitude. But that is really not enough. Feynman rules have a true physics content, and the physicist must understand that. He/she must know how Lorentz invariance, conservation of probability, renormalizability reflect themselves in the Feynman rules. In other words, even if there is no rigorous foundation for these rules, the physical principles at stake must be understood.

This then is the aim: to make it clear which principles are behind the rules, and to define clearly the calculational details. This requires some kind of derivation. The method used is basically the canonical formalism, but anything that is not strictly necessary has been cut out. No one should have an excuse not understanding this book. Knowing about ordinary non-relativistic quantum mechanics and classical relativity one should be able to understand the reasoning.

This book is somewhat unusual in that I have tried very hard to avoid numbering the equations and the figures. This has forced me to keep all derivations and arguments closed in themselves, and the reader needs not to have his fingers at eleven places to follow an argument.

I am indebted to my friends and colleagues R. Akhoury, F. Erné, P. Federbush, P. Van Nieuwenhuizen and F.J. Yndurain. They have read the manuscript critically and suggested many improvements.

The help of M. Jezabek in unraveling the complications of metric usage is gratefully acknowledged. I have some hope that this matter can now finally be put to rest, by providing a very simple translation dictionary.

Ann Arbor, December 1993

1
Lorentz and Poincaré Invariance

1.1 Lorentz Invariance

To begin with we will very briefly review some aspects of Einstein's theory of relativity that are of particular importance here.

The theory of relativity states that physical laws are the same in two systems that move with respect to each other with uniform velocity. Furthermore, the speed of light is a constant. These two statements lead to the concept of invariance under Lorentz transformations. We now first investigate Lorentz transformations in some detail. Some of the rather basic mathematics involved is summarized in an appendix.

Lorentz transformations can be understood as rotations in four-dimensional space (three-dimensional space + time). A rotation can be specified by a matrix L with the property that $\tilde{L}$ (the reflected of L) is its inverse:

$$\tilde{L} = L^{-1} \qquad \text{or} \qquad L\tilde{L} = 1 .$$

Writing indices explicitly:

$$L_{\mu\nu}\tilde{L}_{\nu\lambda} = \delta_{\mu\lambda} \quad \text{or} \quad L_{\mu\nu}L_{\lambda\nu} = \delta_{\mu\lambda} .$$

We have used here Einstein's summation convention: twice occurring indices (such as ν here) are summed over (in this case $\nu = 1, \ldots, 4$). At this point we must also settle some conventions. We take the fourth dimension as imaginary, $x_4 = ict$. This leads to the fact that the matrix-elements of a Lorentz transformation are imaginary if one (but not both) of the indices is four. With this convention a particle at rest has the four-momentum $(0, 0, 0, iM)$, where c has already been taken to be one.

Let us emphasize that there is no physics in the choice of metric. Some physicists prefer to work with real space/time but define their dot-product with a metric involving minus signs. It is really

1

of no relevance where you hide your minus signs, at most it is a matter of convenience. Which is usually what you are used to. It is a matter though that you can debate hotly at lunch time (real time). See appendix on metric.

Examples of Lorentz transformations are:

• ordinary rotations in three dimensions such as a rotation over an angle ϕ around the third axis;

$$L = \begin{bmatrix} \cos\phi & \sin\phi & 0 & 0 \\ -\sin\phi & \cos\phi & 0 & 0 \\ 0 & 0 & 1 & 0 \\ 0 & 0 & 0 & 1 \end{bmatrix}$$

• transformation to a system moving with velocity v along the first axis:

$$L = \begin{bmatrix} \cos\theta & 0 & 0 & \sin\theta \\ 0 & 1 & 0 & 0 \\ 0 & 0 & 1 & 0 \\ -\sin\theta & 0 & 0 & \cos\theta \end{bmatrix}$$

where θ is imaginary and such that $\sin\theta = iv/c\beta$, $\beta = \sqrt{1 - v^2/c^2}$. It follows that $\cos\theta = \sqrt{1 - \sin^2\theta} = 1/\beta$. This transformation is a rotation over an imaginary angle.

In addition to the Lorentz transformations that have determinant 1, such as the ordinary rotations and the velocity transformations there are also transformations with determinant -1. These are the space or time reflections. These are not transformations that you can actually do: nobody has ever managed to reflect himself, transforming himself from, say, a right handed person into a left handed person. In particle physics it has been discovered that the laws of nature are not invariant with respect to these reflections, although large parts of the interactions are. The reflections remain therefore important tools in classifying interactions and establishing selection rules.

A reflection is the combination of any ordinary Lorentz transformation and a space reflection P or time reversal T:

$$P = \begin{bmatrix} -1 & 0 & 0 & 0 \\ 0 & -1 & 0 & 0 \\ 0 & 0 & -1 & 0 \\ 0 & 0 & 0 & 1 \end{bmatrix} \qquad T = \begin{bmatrix} 1 & 0 & 0 & 0 \\ 0 & 1 & 0 & 0 \\ 0 & 0 & 1 & 0 \\ 0 & 0 & 0 & -1 \end{bmatrix}$$

talk in terms of an axis and an imaginary angle also in this case, but that is not important at this moment. The important point is that we observe that a Lorentz transformation is specified by six parameters. Three have a finite, three an infinite domain. The Lorentz-transformations form a six-parameter group.

Since any finite rotation can be seen as an infinite sequence of infinitesimal rotations it is sufficient for most purposes to understand infinitesimal Lorentz transformations. Let us first consider a rotation over an angle ϕ around the third axis. Its form has been given above, and we will denote it by $L(\phi)$. This rotation can be obtained also by applying n times a rotation over an angle ϕ/n:

$$L(\phi) = \left[L\left(\frac{\phi}{n}\right) \right]^n .$$

Let us now consider a rotation over an angle ϕ/n with very large n. We may then expand $\sin(\phi/n)$ and $\cos(\phi/n)$ to get:

$$L(\phi/n) = \begin{bmatrix} 1 & \phi/n & 0 & 0 \\ -\phi/n & 1 & 0 & 0 \\ 0 & 0 & 1 & 0 \\ 0 & 0 & 0 & 1 \end{bmatrix} + \mathcal{O}\left(\phi^2/n^2\right)$$

$$= I + \frac{\phi}{n} L_3 + \mathcal{O}\left(\phi^2/n^2\right)$$

with

$$L_3 = \begin{bmatrix} 0 & 1 & 0 & 0 \\ -1 & 0 & 0 & 0 \\ 0 & 0 & 0 & 0 \\ 0 & 0 & 0 & 0 \end{bmatrix}$$

and I denoting the unit matrix. In the limit of large n:

$$L(\phi) = \lim_{n\to\infty} \left[L\left(\frac{\phi}{n}\right) \right]^n = e^{\phi L_3} .$$

Exercise 1.1 Read the appendix on matrices or else show that

$$\left[1 + \frac{\alpha}{n} + \mathcal{O}\left(\frac{1}{n^2}\right) \right]^n = e^\alpha + \mathcal{O}\left(\frac{1}{n}\right) .$$

We have now written this Lorentz transformation in exponential form. The great advantage is that the parameter ϕ is directly visible, and the property $L(\phi)L(\phi) = L(2\phi)$ is manifest:

$$e^{\phi L_3} e^{\phi L_3} = e^{2\phi L_3}.$$

Similarly other rotations may be treated. A general infinitesimal rotation in three dimensions differs infinitesimally from the unit matrix:

$$R = \begin{bmatrix} 1+g & a & b & 0 \\ d & 1+h & c & 0 \\ e & f & 1+k & 0 \\ 0 & 0 & 0 & 1 \end{bmatrix}$$

with $a \ldots k$ infinitesimal. For a rotation the equation $R\tilde{R} = 1$ holds, and this leads to the result $h = g = k = 0$, $d = -a$, $e = -b$ and $f = -c$ (ignoring higher order terms in a, b, etc.).

Exercise 1.2 Prove this assertion.

We therefore can write:

$$R = I + cL_1 - bL_2 + aL_3$$

with

$$L_1 = \begin{bmatrix} 0 & 0 & 0 & 0 \\ 0 & 0 & 1 & 0 \\ 0 & -1 & 0 & 0 \\ 0 & 0 & 0 & 0 \end{bmatrix} \qquad L_2 = \begin{bmatrix} 0 & 0 & -1 & 0 \\ 0 & 0 & 0 & 0 \\ 1 & 0 & 0 & 0 \\ 0 & 0 & 0 & 0 \end{bmatrix}$$

$$L_3 = \begin{bmatrix} 0 & 1 & 0 & 0 \\ -1 & 0 & 0 & 0 \\ 0 & 0 & 0 & 0 \\ 0 & 0 & 0 & 0 \end{bmatrix}$$

The reason for the sign choices above will become clear shortly. Since a finite transformation can be obtained by exponentiation of an infinitesimal one we so find a representation in terms of three parameters for any rotation in three dimensions:

$$R = e^{\alpha_1 L_1 + \alpha_2 L_2 + \alpha_3 L_3} = e^{\alpha_i L_i}.$$

The three quantities α_i are precisely equal to the vector used to describe rotations introduced above. Thus the direction of α is the axis of rotation, the magnitude is the magnitude of the rotation in radians. The sense of the rotation is this: if $\vec{\alpha}$ points upwards, along the positive third axis, then a small rotation will turn a vector along the positive first axis into a vector having a small negative second component. From the above this connection is obvious for the special cases of rotations around first, second, or third axis. For the general case this becomes obvious by considering an infinitesimal rotation:

$$\begin{bmatrix} 1 & \alpha_3 & -\alpha_2 & 0 \\ -\alpha_3 & 1 & \alpha_1 & 0 \\ \alpha_2 & -\alpha_1 & 1 & 0 \\ 0 & 0 & 0 & 1 \end{bmatrix} , \alpha_i \text{ infinitesimal.}$$

This matrix describes an infinitesimal rotation over an angle $\Delta = \sqrt{\alpha_1^2 + \alpha_2^2 + \alpha_3^2}$ around an axis in the direction $\vec{\alpha}$. The choice of signs for the L_i was made such as to obtain this.

Exercise 1.3 Verify the above by showing that a vector in the direction of $\vec{\alpha}$ is invariant, while a vector perpendicular to $\vec{\alpha}$, for example the unit vector $\vec{r}$ with components $\lambda(\alpha_2, -\alpha_1, 0, 0)$ with

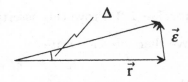

$\lambda = 1/\sqrt{(\alpha_1^2 + \alpha_2^2)}$, is changed by an amount corresponding to a rotation over an angle Δ. Thus compute the effect of the infinitesimal rotation on $\vec{r}$, writing the result in the form $\vec{r} + \vec{\epsilon}$. Show that $\vec{\epsilon}$ is orthogonal to $\vec{r}$ and $\vec{\alpha}$, and has the magnitude Δ.

This treatment can be extended trivially to include the "velocity" transformations. A general "velocity" transformation will be of the form:

$$V = e^{\beta_1 M_1 + \beta_2 M_2 + \beta_3 M_3}$$

with imaginary β_1, β_2 and β_3 and real M_1, M_2 and M_3:

$$M_1 = \begin{bmatrix} 0 & 0 & 0 & 1 \\ 0 & 0 & 0 & 0 \\ 0 & 0 & 0 & 0 \\ -1 & 0 & 0 & 0 \end{bmatrix} \qquad M_2 = \begin{bmatrix} 0 & 0 & 0 & 0 \\ 0 & 0 & 0 & 1 \\ 0 & 0 & 0 & 0 \\ 0 & -1 & 0 & 0 \end{bmatrix}$$

$$M_3 = \begin{bmatrix} 0 & 0 & 0 & 0 \\ 0 & 0 & 0 & 0 \\ 0 & 0 & 0 & 1 \\ 0 & 0 & -1 & 0 \end{bmatrix}$$

Of course we could equally well have used real $\vec{\beta}$ and imaginary

M, but we will do it this way. Writing $\vec{\beta} = i\vec{v}$, a particle of mass m at rest is transformed into a particle moving with momentum $\vec{p} = -m\vec{v}$. Note that the $\vec{v}$ here is the conventional velocity divided by the relativistic factor β.

The general Lorentz transformation is of the form

$$L = e^{\alpha_i L_i + \beta_i M_i}$$

but the interpretation of the α_i and β_i in terms of axis of rotation or velocity is no more as easy as above.

At this point it is necessary to introduce a new and sometimes slightly confusing notation. We write:

$$L = e^{\frac{1}{2}\alpha_{\mu\nu}K_{\mu\nu}}, \quad \mu, \nu = 1, \ldots 4.$$

The matrices K are defined by the prescription that $K_{\mu\nu}$ is a matrix with 1 in row μ, column ν, and -1 in row ν, column μ. Otherwise its elements are zero. Note that $K_{\mu\nu} = -K_{\nu\mu}$. The $\alpha_{\mu\nu}$ are chosen such as to give the correct result. Thus, given that $L_1 = K_{23}$, $L_2 = -K_{13}$ and $L_3 = K_{12}$ the correspondence is:

$$\alpha_1 \leftrightarrow \alpha_{23} \qquad \beta_1 \leftrightarrow \alpha_{14}$$
$$\alpha_2 \leftrightarrow \alpha_{31} \qquad \beta_2 \leftrightarrow \alpha_{24}$$
$$\alpha_3 \leftrightarrow \alpha_{12} \qquad \beta_3 \leftrightarrow \alpha_{34}$$

while the remaining α are defined by $\alpha_{\mu\nu} = -\alpha_{\nu\mu}$.

The confusion may arise by not being careful about indices. The K are 4×4 matrices, the α are numbers with $\alpha_{\mu\nu}$ real if $\mu, \nu = 1, 2, 3$, or $\mu = \nu = 4$, and imaginary if μ or $\nu = 4$. To be very explicit, the matrix-element i, j of the matrix K_{13} could be written as

$$(K_{13})_{ij}.$$

1.2 Structure of the Lorentz Group

We must now study the structure of the Lorentz group, by which we mean the following. Two successive Lorentz transformations equals another Lorentz transformation, and we must understand this connection in terms of the parameters $\alpha_{\mu\nu}$. Thus, let there be given two Lorentz transformations described by parameters $\alpha_{\mu\nu}$ and $\beta_{\mu\nu}$ respectively. The product of these two is another Lorentz transformation described by parameters $\gamma_{\mu\nu}$ and we would like to know the γ as function of the α and β. Unfortunately this relation is quite complicated. On the infinitesimal level it is relatively easy to compute γ, but the complete expression is more difficult.

The starting equation is:

$$L(\gamma) = L(\beta)\, L(\alpha).$$

In lowest order:

$$\gamma_{ij} = \beta_{ij} + \alpha_{ij} + \text{ terms of higher order.}$$

It is not that difficult to go to the next order. First consider the product

$$e^A\, e^B,$$

where A and B are matrices, to second order in A and B.

$$\left(1 + A + \tfrac{1}{2}A^2\right)\left(1 + B + \tfrac{1}{2}B^2\right) \simeq 1 + A + B + AB + \tfrac{1}{2}B^2 + \tfrac{1}{2}A^2 + \ldots$$

Note that AB need not to be equal to BA. It is easy to see that up to higher order terms this is equal to

$$1 + C + \tfrac{1}{2}C^2$$

with $C = A + B + \tfrac{1}{2}AB - \tfrac{1}{2}BA = A + B + \tfrac{1}{2}[A,B]$. Thus, writing

$$e^C = e^A e^B$$

we have up to second order in A and B:

$$C = A + B + \tfrac{1}{2}[A,B] + \ldots$$

with $[A,B] = AB - BA$. In the same way subsequent terms of C may be worked out:

$$C = A + B + \tfrac{1}{2}[A,B] + \tfrac{1}{12}[A,[A,B]]$$
$$+ \tfrac{1}{12}[[A,B],B] + \tfrac{1}{48}[A,[[A,B],B]] + \ldots$$

involving multiple commutators such as

$$[[A,B],B]] = [A,B]B - B[A,B]$$
$$= ABB - 2BAB + BBA.$$

It may be proven that all higher order terms can be written as multiple commutators. The equation is called the **Campbell–Baker–Hausdorff** formula, and in the appendix on matrices this assertion is proven. We do not need the explicit form, but only the fact that all terms are multiple commutators.

To work out the multiplication laws for the Lorentz transformations we must work out multiple commutators for the matrices K.

Here it turns out that the commutator of any two K matrices is a linear combination of K matrices. This is really a consequence of the fact that the transformations form a group, the product of two transformations is again a Lorentz transformation and can also be expressed as an exponential involving K matrices. Thus we need to work out only the commutators of any pair of K matrices. That is still some work, but not too much. Things are somewhat easier in terms of the original matrices L_i and M_i, for example

$$[L_1, L_2] = -L_3$$

and cyclically. Also the commutators between the M and the L and M are easy to do. Nevertheless, writing everything in terms of the K leads to a formidable expression:

$$[K_{ab}, K_{cd}] = c^{ij}_{abcd} \, K_{ij}$$

(summation over i and j from 1...4) with the coefficients c given by

$$c^{ij}_{abcd} = \delta_{bc}\delta_{ai}\delta_{dj} - \delta_{bd}\delta_{ai}\delta_{cj} - \delta_{ac}\delta_{bi}\delta_{dj} + \delta_{ad}\delta_{bi}\delta_{cj}$$

Exercise 1.4 Check this equation.

For example, for $a = 2$, $b = 3$, $c = 3$, $d = 1$ we have

$$c^{ij}_{2331} = \delta_{2i}\delta_{1j}$$

thus

$$[K_{23}, K_{31}] = K_{21} = -K_{12}$$

which is the same as $[L_1, L_2] = -L_3$.

Since all terms in the Campbell–Baker–Hausdorff formula are given in terms of multiple commutators we can express everything in terms of the coefficients c. For example, going back to our original equation:

$$\gamma^{ij} = \beta^{ij} + \alpha^{ij} + \tfrac{1}{2} c^{ij}_{abcd}\beta^{ab}\alpha^{cd} + \tfrac{1}{12}cc\beta\beta\alpha + \dots$$

Summation over a, b, c, and d from 1 to 4 is understood. In the last term we did not write all indices, but just indicated which quantities are involved. A certain symmetry in notation has been achieved by putting the indices on the α, β and γ on top.

The constants c are called structure constants since knowledge of the c implies knowledge of the structure of the Lorentz group.

They are the central quantities of the group. Knowing the c one knows everything about products of Lorentz transformations. One can say that knowing a group amounts to knowing its structure constants.

1.3 Poincaré Invariance

Laws of nature seem not only invariant with respect to Lorentz invariance, they also appear to be invariant under translations. Thus Maxwell's laws appear equally valid in Europe or the U.S. (translation in space) or yesterday or today (translation in time). We observe thus invariance under the transformation

$$\vec{x} \rightarrow \vec{x} + \vec{b}$$
$$t \rightarrow t + b_0$$

or

$$x \rightarrow x + b$$

where now x and b denote 4-vectors. The combined invariance under Lorentz transformations and translations is called **Poincaré invariance**. It should be noted that Lorentz transformations and translations do not commute (first a translation $T(b)$ over a vector b followed by a Lorentz transformation L is in general not the same as L followed by $T(b)$).

1.4 Maxwell Equations

The Maxwell equations describe electricity and magnetism, and also light, and we must face the problem of quantizing these equations. We know that light is quantized in photons, thus since light is nothing but a combination of electric and magnetic fields we conclude that these fields are quantized. As it happens the Maxwell equations, while well understood classically, are quite difficult in the context of quantum field theory. The problem is with gauge invariance. The classical equations for the potentials are:

$$\partial_\mu F_{\mu\nu} = j_\nu$$

with

$$F_{\mu\nu} = \partial_\mu A_\nu - \partial_\nu A_\mu \,,$$

where the four-vector j_μ describes currents and charges. The classical expression for the energy density (energy/unit volume) in terms of the electric and magnetic fields is:

$$\mathcal{H} = \tfrac{1}{2}\left[\vec{H}^2 + \vec{E}^2\right] = \tfrac{1}{4}F_{\mu\nu}F_{\mu\nu} + F_{4\mu}F_{\mu4}\,,$$

$$H_1 = F_{23}\,, \quad H_2 = -F_{13}\,, \quad H_3 = F_{12}\,, \quad E_j = iF_{j4}\,, \quad j = 1, 2, 3.$$

Exercise 1.5 Verify this.

If now some potential $A_\mu(x)$ is a solution of the equations then also

$$A_\mu + \partial_\mu\Lambda(x)$$

with Λ any differentiable function, is a solution with the same energy density. How can this work quantum mechanically?

If light is made up from photons then the energy density of light at some point at some moment is the probability of finding photons at that point at that moment times the energy of the photons. Now classically, the energy density is proportional to A_μ^2 (for free photons), thus apparently in quantum mechanics the potential becomes something like the photon wave function. Indeed, the square of a wave function gives the probability distribution, which when multiplied by the energy/photon is the energy density.

Now there are four components to the vector potential and there are thus four different kinds of photons, each with their own wave function. They may be transformed into each other by a rotation, or more general a Lorentz transformation, that as we know transforms the various components of the A_μ into each other. Gauge invariance tells us that adding something to these wave functions is physically of no relevance. This leads us to the conclusion that in the description of the e.m. field there will be photon types that cannot be observed, that do not interact with matter, but seem to be part of the formalism. This is indeed what happens and that makes the quantum theory of electromagnetism quite difficult.

There are two ways of coping with this problem. First, we may go ahead and indeed introduce the unphysical photons next to the physical ones. There are two of the latter, corresponding to the two polarization states of light. Thus there are then two unphysical types, which in fact one must give quite unphysical properties (such as negative probability) to obtain an elegant mathematical description.

The other way is to slightly modify Maxwell's equations by giving the photons a very, very small mass. This corresponds to the equation:

$$\partial_\mu \left(\partial_\mu A_\nu - \partial_\nu A_\mu \right) - \kappa^2 A_\nu = j_\nu$$

where κ is the photon mass. Now gauge invariance has disappeared, because the mass term is not invariant. If the current is conserved ($\partial_\mu j_\mu = 0$) then by taking the derivative of the above equation we find

$$\partial_\nu A_\nu = 0$$

which is a relation between four components of the A_ν. Thus in that case there will be three independent "photons", and we need no unphysical photons. Again we have a problem, because light as we know it has two states and not three. As it happens, one can prove that in this theory in the limit $\kappa \to 0$ one of the "photons" decouples and one recovers Maxwell's theory.

No matter how we do it, it seems that we must introduce extra particles. We will take the approach of giving a mass to the photon, because we do not want to deal too quickly with unphysical particles.

We will come back to these interesting questions in more detail at the appropriate place.

1.5 Notations and Conventions

A physical space-time point $(\vec{x}, t)$ is described by a four component vector x_ν, or x for short, with $x_4 = ict$. We will choose units such that c, the speed of light, is one, while also $\hbar$ is 1. This leaves one free unit for which we take the unit of energy MeV. In these units for example the electron mass is 0.511 MeV. From $E = mc^2$ we see that energy and mass are the same in this system. If $\hbar$ is to be one (in conventional units $\hbar = 6.582122 \times 10^{-22}$ MeV sec) as well as $\hbar c$ ($= 1.97327 \times 10^{-11}$ MeV cm) then this defines units of length and time. As a consequence, if a calculation of a length is done the answer will be in 1/MeV and conversion to cm follows by multiplying with $\hbar c$ as quoted above. Similarly for time.

Integration over space and time is over real space, real time. Thus

$$\int d_4x = \int d_3x \, dx_0.$$

A particle of mass m and momentum $\vec{p}$ has an energy E given by

$$E = \sqrt{\vec{p}^2 + m^2}.$$

This energy E is essentially the fourth component of p:

$$p_\mu = (\vec{p}, iE).$$

We will often use the notation x_0 or p_0 by which we mean the fourth component without the i. Thus $x_4 = ix_0$, $p_4 = ip_0 = iE$.

The energy–momentum relation may be plotted in a graph where the horizontal axis represents $\vec{p}$ and the vertical p_0. The relation $p_0^2 - \vec{p}^2 = m^2$ gives two curves, one with positive and one with negative p_0. For zero mass these curves become a cone, the light cone. Since the horizontal axis represents in fact a three-dimensional space these curves and cones represent surfaces. These surfaces are called mass-shells.

The energy–momentum relation $p_0^2 - \vec{p}^2 = m^2$, or $(p \cdot p) = -m^2$ (with $p = \vec{p}, ip_0$) is clearly invariant under Lorentz transformations. Thus a vector on a given mass-shell remains on that mass-shell under a Lorentz transformation.

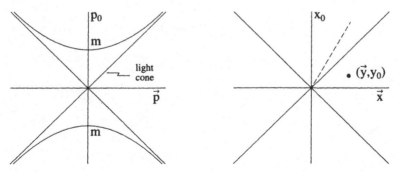

In space-time $x = (\vec{x}, x_0)$ we may plot the motion of a particle. If the particle moves with constant speed we have a straight line. Since no particle can go faster than the speed of light the angle of these lines with the x_0 axis cannot be less than $45°$. The $45°$ lines describe motion of light. Again, we have a light cone. Particles remain within that cone. A space-time point $\vec{y}, y_0$ outside the light cone cannot ever receive information of an event that took place at time 0 at the point $\vec{0}$ because speeds exceeding that of light would be required.

Under a Lorentz transformation dot-products such as $(y \cdot y) = \vec{y}^2 + y_4^2 = \vec{y}^2 - y_0^2$ remain invariant.

Exercise 1.6 Make a plot of all possible points that $\vec{y}, y_0$ can be transformed to by a Lorentz transformation. Next, take any such point and show that there is actually a Lorentz transformation that transforms $\vec{y}, y_0$ into that point.

2
Relativistic Quantum Mechanics of Free Particles

2.1 Hilbert Space

In a quantum mechanical description the state of a free spinless particle is completely specified by its three-momentum. The energy follows from the energy–momentum relation. Location in space and time is completely unknown. Of course, we could also specify location $\vec{x}$, but then the momentum would be unknown. But first we concentrate on the momentum description. Also we will start with a particle of which only one type exists, unlike the photon or a massive photon. Such particles are called **scalars** or **pseudoscalars** depending on the behaviour of the wave function under reflection. The π^0 is an example of such a particle, it is a pseudoscalar.

Let us pause a moment to reflect on the physics of the above statements. In conventional presentations of quantum mechanics one encounters a postulate, prescribing the commutation relation between momentum and location operators. The essential physical content of that mathematical statement is this: a particle with well defined momentum (and energy) is described by a plane wave. This is the true content of the Planck–Einstein relation $E = h\nu$. It could have been different. A particle of well defined energy could have corresponded to some other wave type. But this is the idea: a particle of well defined energy is a plane wave with frequency ν. Thus we do not postulate commutation rules, but we postulate that a particle of well defined momentum and energy is described by a plane wave.

The additional quantum mechanical concept is superposition. It is borrowed from the wave theory of light. If we say that light is made up from photons, then necessarily we must identify the intensity of light (which is proportional to the square of the

15

amplitude) at some point with the probability to detect a photon at that point. A wave packet, with reasonably confined intensity, must describe a reasonably confined photon. These wave packets, in the classical theory of light, are superpositions of plane waves. A generalization of quantum mechanics is that waves may be complex and that the probability is not the square, but the absolute value squared of the amplitude. What follows below is the mathematical implementation of these statements.

We assume the existence of an infinite dimensional space, Hilbert space, and to every possible physical state corresponds a vector in Hilbert space. Vectors are orthogonal if the corresponding states are mutually exclusive. For example, if a particle has momentum $\vec{p}$ then the probability of finding $\vec{q}$ (with $\vec{q}$ different from $\vec{p}$) when measuring the momentum is zero. The vectors corresponding to particles with momenta $\vec{p}$ and $\vec{q}$ ($\vec{p} \neq \vec{q}$) are thus orthogonal. We choose these vectors to be basis vectors of length 1 and denote them by $\mid \vec{p} >$ and $\mid \vec{q} >$. In this Hilbert space we have the following basis vectors:

$\mid 0 >$ corresponding to the vacuum

$\mid \vec{p}_1 >$ corresponding to a particle of momentum $\vec{p}_1$

$\mid \vec{p}_2 >$ corresponding to a particle of momentum $\vec{p}_2$

$\vdots$ $\vdots$

$\mid \vec{p}_1, \vec{q}_1 >$ corresponding to two particles, one of momentum $\vec{p}_1$, the other $\vec{q}_1$

$\vdots$

$\mid \vec{p}_1, \vec{q}_1, \vec{k}_1 >$

$\vdots$

    etc.

Let us repeat, the states are normalized to one. For example, the dot-product between the states $\mid \vec{p} >$ and $\mid \vec{q} >$ will be denoted by $< \vec{p} \mid \vec{q} >$ and is equal to $\delta_{\vec{p}\vec{q}}$. Since the range of momenta is infinite the space of the single particles above is already of infinite dimension. There is a problem because the momentum range is continuous. This is usually considered a formal problem that can be resolved as follows. Assume that the universe is a square box with volume V. Allow only wave functions whose value on the boundary is the same as the value on the opposite boundary (periodic boundary conditions). This restricts momenta to the values (L = length of side of the cube)

$$p_1 = \pm \frac{2\pi\, n_1}{L}, \ p_2 = \pm \frac{2\pi\, n_2}{L}, \ p_3 = \pm \frac{2\pi\, n_3}{L}, \quad n_1, n_2, n_3 = 0, 1, \ldots$$

Now the momenta are discrete. In any calculation we may at the end take the limit $V \to \infty$, thereby going to the continuous case. Then a sum over momenta becomes an integral:

$$\sum_{\vec{p}} \to \frac{V}{(2\pi)^3} \int d_3 p.$$

The factor $V/(2\pi)^3$ arises because in a little box $d_3 p$ there will be that many possible momenta.

Exercise 2.1 Check this. Also verify the formal equation

$$\lim_{V \to \infty} \frac{V}{(2\pi)^3} \delta_{\vec{p}_i \vec{p}_j} = \delta_3 \left(\vec{p}_i - \vec{p}_j \right)$$

The equation is formal, and makes sense only if there is a summation (integration). Then it means:

$$\sum_{\vec{p}_i} \delta_{\vec{p}_i \vec{p}_j} f(\vec{p}_i) = f(\vec{p}_j) = \int d_3 p_i\, \delta_3(\vec{p}_i - \vec{p}_j) f(\vec{p}_i).$$

What we are doing here is of course quite horrible: we are violating Lorentz invariance. A square box of volume V is not Lorentz invariant, it is only invariant for some very special spatial transformations, such as a rotation over 90 degrees around the third axis. At this point we may only keep our fingers crossed and hope that the results will be Lorentz invariant in the limit of infinite V. This plague, having to abandon Lorentz invariance in order to define the formalism, seems common to all approaches to quantum field theory. One always needs some kind of a grid. The final results, the Feynman rules, do not suffer from this breaking of Lorentz invariance. We will, by necessity, ignore this difficult issue.

We now must review some of the basic concepts of quantum

mechanics. The wave function of a non-relativistic particle is:

$$\psi(x,t) = \frac{1}{\sqrt{V}} e^{i(\vec{p}\vec{x} - E_k t)}$$

E_k is the non-relativistic kinetic energy,

$$E_k = \tfrac{1}{2} m \vec{v}^{\,2} = \frac{\vec{p}^{\,2}}{2m}.$$

The normalization is such that the total probability (integration over the whole universe of volume V) is 1. This describes one particle in the universe.

The above wave function is a solution of the Schrödinger equation for a free particle:

$$\frac{1}{2m} \Delta \psi = -i \frac{\partial \psi}{\partial t}.$$

The relativistic generalization of the wave function is simple:

$$\psi(x,t) = \frac{1}{\sqrt{V}} e^{i(px)} = \frac{1}{\sqrt{V}} e^{i(\vec{p}\,\vec{x} - Et)}$$

where now E is the full relativistic expression for the energy including the rest-mass energy:

$$E = \sqrt{\vec{p}^{\,2} + m^2}.$$

This wave function now satisfies the Klein–Gordon equation:

$$\left(\Box - m^2 \right) \psi(x) = 0, \quad \text{with} \quad \Box = \frac{\partial^2}{\partial x_1^2} + \frac{\partial^2}{\partial x_2^2} + \frac{\partial^2}{\partial x_3^2} + \frac{\partial^2}{\partial x_4^2}.$$

Exercise 2.2 Verify this.

However, the definition of probability needs some revision. In the non-relativistic theory the probability density to find a particle at some point x within a small box $d_3 x$ is given by

$$|\psi(x)|^2 d_3 x.$$

The integral of this expression over the whole volume gives 1. However, since volume is not a Lorentz invariant quantity (Lorentz contraction of length) we cannot maintain this definition of probability density. Probability density is not Lorentz invariant, only the total probability must be Lorentz invariant (and is equal to 1). This is very much like electric charge. In fact, thinking of an electron, charge density must be proportional to probability density. This then leads us to introducing probability current and probability density, with a conservation equation:

$$\partial_\mu P_\mu(x) = 0$$

analogous to the conservation equation of electric charge. Likewise, from this equation one can prove that the total probability:

$$P = \int d_3x P_0(x)$$

is a constant $(\partial P/\partial t = 0)$, using that the probability current P_μ, $\mu = 1, 2, 3$ is zero on the boundary.

Exercise 2.3 Verify this.

A definition of probability current and density satisfying all these requirements, and which is moreover such that the non-relativistic expression follows for P_0 in the low energy limit is:

$$P_\mu(x) = i\frac{\partial \psi^*(x)}{\partial x_\mu}\psi(x) - i\psi^*(x)\frac{\partial \psi(x)}{\partial x_\mu}.$$

In the low energy limit $E \approx m$ and $P_0 \approx 2m\psi^*\psi$. By normalizing the ψ as follows

$$\psi(x) = \frac{1}{\sqrt{2Vp_0}}e^{i(px)}$$

we have normalization to 1 when integrating P_0 over the whole volume V.

Exercise 2.4 Show that $\partial_\mu P_\mu(x) = 0$.

Let us now consider other than pure plane wave states. A particle of which we know that it is precisely at the point $\vec{x}$ at some time x_0 (we will take $x_0 = 0$) is described at time 0 by a δ-function:

$$\psi(\vec{x}, 0) = C\delta_3(\vec{x})$$

where C is a normalization constant*. This can be seen as a superposition of pure momentum states:

$$\psi(\vec{x}, 0) = \frac{C}{(2\pi)^3}\int d_3p\, e^{i\vec{p}\vec{x}}$$

---

* Actually, this wave function, a δ-function, can not be normalized. Strictly speaking one should work only with wave packets, slightly smeared out.

from which we generalize:

$$\psi(\vec{x}, t) = \psi(x) = \frac{C}{(2\pi)^3} \int d_3 p \, e^{i(px)}$$

with now $(px) = \vec{p}\vec{x} + p_4 x_4 = \vec{p}\vec{x} - Et$, with $E^2 = \vec{p}^2 + m^2$. This generalization is such that the wave function satisfies the Klein–Gordon equation.

If the location is not a δ-function but more smeared out then we have, in general:

$$\psi(x) = \frac{1}{(2\pi)^3} \int d_3 p \, f(p) e^{i(px)}, \qquad p_0^2 = \vec{p}^2 + m^2$$

with $f(p)$ some function of the momentum p. If $f(p) = 1$ we have a δ-function in space, if $f(p) = \delta_3(\vec{p} - \vec{q})$ we have a state with sharply defined momentum $\vec{q}$.

From this we may deduce the vector in Hilbert space that corresponds to a sharply defined location at some time. It will be a superposition of sharp momentum states with equal weight:

$$\mid \vec{x}, x_0 > \; = \sum_{\vec{p}} C e^{i(px)} \mid \vec{p} > .$$

where C is some constant such that normalization holds for $\mid \vec{x}, x_0 >$, i.e., that $\mid \vec{x}, x_0 >$ is a vector of unit length. This requires that we define length for a vector in Hilbert space, which we will do below.

From the above treatment we learn a very simple and important fact. In general, if we do a Lorentz or Poincaré transformation then a state in Hilbert space will transform to another state. For instance, if a Lorentz transformation L transforms the momentum p into p', i.e., $Lp = p'$, then we evidently must also transform Hilbert space such that the vector $\mid \vec{p} >$ becomes the vector $\mid \vec{p}' >$. Similarly two and more particle vectors transform to other vectors in Hilbert space. Thus for example $\mid \vec{p}, \vec{q} >$ becomes $\mid \vec{p}', \vec{q}' >$ where p' and q' are the Lorentz transforms of p and q. Thus corresponding to a Lorentz transformation there is a complicated, big transformation in Hilbert space. To describe it needs an ∞ by ∞ matrix. Fortunately, we will not need the explicit form of that matrix, because it is very simple to say what it does. It must, however, be realized that the Lorentz transformation matrix in Hilbert space is very different from the 4×4

matrix describing the Lorentz transformation in four-dimensional space.

Needless to say that the above makes sense only in the limit of infinite V, when there is some hope of having a Lorentz invariant situation. Indeed, for finite V, when the momenta are discrete, the Lorentz transform of a momentum will in general not precisely fall onto one of the allowed discrete values. But again, we will ignore this issue.

The behaviour under translations is much simpler and we will discuss that now. Under a translation b a particle sharply located at place $\vec{x}$, time x_0, will become a particle sharply located at place $\vec{x} + \vec{b}$, time $x_0 + b_0$. Thus:

$$T(b) \quad : \quad \mid x > \rightarrow \mid x + b > .$$

Thus:

$$\sum_{\vec{p}} e^{i(px)} \mid \vec{p} > \rightarrow \sum_{\vec{p}} e^{i(px)+i(pb)} \mid \vec{p} > .$$

Under a translation the state $\mid \vec{p} >$ goes over into the state

$$e^{i(pb)} \mid \vec{p} >$$

which shows explicitly how the transformation of the vectors in Hilbert space differs from what happens in ordinary space. Note that the vector $\mid \vec{p} >$ gets just a phase factor, in other words, it stays normalized and represents, physically speaking, the same state.

The conclusion of the above is the following extremely important statement:

> **To every Lorentz transformation, or more generally a Poincaré transformation, corresponds a transformation in Hilbert space.**

Since we want physics to be unique we will insist that this is a one–one correspondence. Later we will relax this requirement slightly allowing a phase factor (for example to a Lorentz transformation may correspond two transformations in Hilbert space that differ only by a phase factor $e^{i\lambda}$).

If the correspondence is unique then to the product of two Lorentz transformations L_1 and L_2 must correspond the product of the two corresponding transformations X_1 and X_2 in Hilbert space. Thus, if

$$L_1 \leftrightarrow X_1$$
$$L_2 \leftrightarrow X_2$$

and to $L_3 = L_1 \cdot L_2$ corresponds X_3, then we must require that $X_3 = X_1 \cdot X_2$.

Exercise 2.5 Check that this property holds for translations, with the appropriate definition for what is the "product" of two translations (the effect of two subsequent translations).

We say that the transformations X in Hilbert space are a representation of the Lorentz group. Again, the matrices X are in general very different from the matrices L.

2.2 Matrices in Hilbert Space

In order to describe physical processes we must introduce operators (matrices) in the previously described Hilbert space. A typical process, for example electrons scattering off a proton, involves different states from the point of view of Hilbert space. Thus initially we may have an electron of momentum $\vec{p}$ and a proton of momentum $\vec{k}$, to which thus corresponds the basis vector $\mid \vec{p}, \vec{k} >$ in Hilbert space and the final state may then contain an electron and proton of momenta $\vec{p}\,'$ and $\vec{k}\,'$ corresponding to the vector $\mid \vec{p}\,', \vec{k}\,' >$ in Hilbert space. One might think of this as if the physical system corresponds to a vector in Hilbert space that rotates as a function of time from $\mid \vec{p}, \vec{k} >$ to $\mid \vec{p}\,', \vec{k}\,' >$. This is **not** the way we are going to describe things. The reason is that scattering clearly involves interaction between particles, and we have set up our Hilbert space for free particles. We must rethink our procedure when we introduce interactions.

Right now it suffices to say that physical quantities will correspond to the elements of a certain matrix defined in Hilbert space. What we need is basic building blocks, in some sense matrices like the Pauli spin matrices that can be used to describe the full set of 2×2 matrices. The matrices that we need will fulfil certain basic requirements, in particular the requirement of locality. This is the requirement that physical processes cannot influence each other if they are outside each other's light cone, i.e., if speeds larger than that of light are needed to connect the events. This we hope will be achieved by insisting that the operator (matrix) describing a process at the space-time point x will commute with a similar operator for the space-time point y if x and y are outside each other's light cone. Locality is a touchy point of quantum theory, and to a large extent it remains, in its formulation, an article of faith. There is however, at this time, no reason to suspect anything wrong with the procedure adopted.

To explain what we are going to do consider first three-dimensional space. What would be the most elementary building blocks that can be used to build up any 3×3 matrix? The claim is that we can do this with two matrices:

$$a = \begin{bmatrix} 0 & 1 & 0 \\ 0 & 0 & \sqrt{2} \\ 0 & 0 & 0 \end{bmatrix} \qquad \tilde{a} = \begin{bmatrix} 0 & 0 & 0 \\ 1 & 0 & 0 \\ 0 & \sqrt{2} & 0 \end{bmatrix}$$

where $\tilde{a}$ is the reflected of a.

Exercise 2.6 Work out the products a^2, $\tilde{a}^2$, $\tilde{a}a$, $a\tilde{a}$, $\tilde{a}a - a\tilde{a}$ etc., and show that any 3×3 matrix can be obtained as a linear combination of these.

The reason that we introduced the factor $\sqrt{2}$ will become more transparent later. We may extend the example to many more dimensions. In a large dimensional space we still can do with two matrices a and $\tilde{a}$ with

$$a = \begin{bmatrix} 0 & 1 & 0 & 0 & 0 & \cdots \\ 0 & 0 & \sqrt{2} & 0 & 0 & \cdots \\ 0 & 0 & 0 & \sqrt{3} & 0 & \cdots \\ 0 & 0 & 0 & 0 & \sqrt{4} & \cdots \\ \vdots & & & & & \end{bmatrix}$$

Exercise 2.7 Show that a applied to the third (counting $0, 1, \ldots$) unit vector gives $\sqrt{3}$ times the second unit vector. Find a^2, $\tilde{a}a^2$ and $\tilde{a}a - a\tilde{a}$. Try to understand how the matrix $\tilde{a}^{m-1}a^m$ would look. Do not try to find the precise numbers, but just which elements would be non-zero.

We now turn to Hilbert space. The above example can be used for the subspace referring to a definite momentum $\vec{p}$. A matrix a like above can be constructed in the subspace of states $\mid 0 >$, $\mid \vec{p} >$, $\mid \vec{p}, \vec{p} >$ etc. The matrix a when applied to a state vector with n particles of momentum $\vec{p}$ gives then the vector for a state with $n-1$ particles of momentum $\vec{p}$ with a factor $\sqrt{n}$. Examples:

$$a \mid \vec{p}, \vec{p}, \vec{p} > \; = \sqrt{3} \mid \vec{p}, \vec{p} > \qquad a \mid 0 > \; = 0.$$

This particular matrix operating in the $\vec{p}$ subspace will be denoted by $a(\vec{p})$. Similarly we will have matrices for any other subspace, and we have then an infinite set of matrices $a(\vec{p})$ and $a^{\dagger}(\vec{p})$ (with $a^{\dagger} = \tilde{a}^{*}$, which is the same as $\tilde{a}$ since a is real here).

Having done this we do not need to introduce new matrices for that part of Hilbert space where we have particles of different momenta, such as the state $\mid \vec{p}, \vec{p}, \vec{q}, \vec{q}, \vec{q} >$ with $\vec{p} \neq \vec{q}$. The matrices $a(\vec{p})$ and $a(\vec{q})$ will by definition act on these states as if the other particles are not there. Thus (with $\vec{p} \neq \vec{q}$):

$$a(\vec{p}) \mid \vec{q}, \vec{q}, \vec{q}, \vec{q}, \vec{q} > \; = 0$$
$$a(\vec{p}) \mid \vec{p}, \vec{p}, \vec{q}, \vec{q}, \vec{q} > \; = \sqrt{2} \mid \vec{p}, \vec{q}, \vec{q}, \vec{q} >$$
$$a(\vec{q}) \mid \vec{p}, \vec{p}, \vec{q}, \vec{q}, \vec{q} > \; = \sqrt{3} \mid \vec{p}, \vec{p}, \vec{q}, \vec{q} >$$

We now have a set of matrices $a(\vec{p})$ and $a^{\dagger}(\vec{p})$, called annihilation and creation operators respectively, defined over the whole Hilbert space, and that can be used to build up any other matrix. Note that thus by construction $a(\vec{p})\, a(\vec{q}) = a(\vec{q})a(\vec{p})$ if $\vec{p} \neq \vec{q}$ and of course also if $\vec{p} = \vec{q}$. They do not interfere with each other.

Exercise 2.8 Show that the matrix

$$H = \sum_{\vec{p}} p_0 a^{\dagger}(\vec{p})a(\vec{p}), \qquad p_0 = \sqrt{\vec{p}^2 + m^2}$$

is diagonal, and moreover it is the energy operator, that is:

$$H \mid \alpha > \; = E_{\alpha} \mid \alpha >$$

where E_{α} is the total energy of the state $\mid \alpha >$. In here α may be any number of particles of any momentum. Start by showing that $H \mid 0 > \; = 0$, $H \mid \vec{q} > \; = q_0 \mid \vec{q} >$, $H \mid \vec{p}, \vec{p} > \; = 2p_0 \mid \vec{p}, \vec{p} >$, $H \mid \vec{p}, \vec{q} > \; = (p_0 + q_0) \mid \vec{p}, \vec{q} >$, etc.

2.3 Fields

The Fourier transforms of the matrices a and $a^\dagger$ are called **fields**. To be precise we have the field $A(x)$:

$$A(x) = \sum_{\vec{p}} \frac{1}{\sqrt{2Vp_0}} a(\vec{p}) e^{ipx},$$

with as usual $px = \vec{p}\vec{x} + p_4 x_4 = \vec{p}\vec{x} - p_0 x_0$ and $p_0 = \sqrt{\vec{p}^2 + m^2}$. Similarly

$$A^\dagger(x) = \sum_{\vec{p}} \frac{1}{\sqrt{2Vp_0}} a^\dagger(\vec{p}) e^{-ipx}.$$

Notice that $A(x)$ is no more a real matrix due to the complex factors $\exp(ipx)$. Also the matrices $A(x)$ and $A^\dagger(x)$ can be taken as basic building blocks. If nothing else, the a can be recovered from the A. For example $a(\vec{q})$ can be obtained from $A(x)$ by another Fourier transformation:

$$\int d_3 x e^{-iqx} A(x) = \sqrt{\frac{V}{2q_0}} a(\vec{q}).$$

This needs some fast footwork concerning the continuum limit, but there is no point in wasting our time here. The hermitian combination

$$\phi(x) = A(x) + A^\dagger(x)$$

is called the field corresponding to the particles considered. This field has a number of properties that make it very useful to construct physical quantities. The main property is that it is local. The commutator

$$[\phi(x), \phi(y)] = \phi(x)\phi(y) - \phi(y)\phi(x)$$

is zero if x and y are outside of each other's light cone. To see this we compute things step by step:

• $[A(x), A(y)] = 0$.
This follows because $a(\vec{p})a(\vec{q}) = a(\vec{q})a(\vec{p})$.
• $\left[A^\dagger(x), A^\dagger(y)\right] = 0$
From $a^\dagger(\vec{p})a^\dagger(\vec{q}) = a^\dagger(\vec{q})a^\dagger(\vec{p})$.
• $\left[A(x), A^\dagger(y)\right] = \sum_{\vec{p}} \frac{1}{2Vp_0} e^{ip(x-y)}$

- $\left[A^\dagger(x), A(y)\right] = -\sum_{\vec{p}} \frac{1}{2Vp_0} e^{-ip(x-y)}$

These last two follow from the equations

$$\left[a(\vec{p}), a^\dagger(\vec{q})\right] = \delta_{\vec{p},\vec{q}} = \begin{cases} 1 & \text{if } \vec{p} = \vec{q} \\ 0 & \text{otherwise} \end{cases}$$

$$\left[a^\dagger(p), a(\vec{q})\right] = -\delta_{\vec{p},\vec{q}}$$

Altogether we find:

$$[\phi(x), \phi(y)] = \sum_{\vec{p}} \frac{1}{2Vp_0} \left\{ e^{ip(x-y)} - e^{-ip(x-y)} \right\}.$$

We now proceed to show that the right hand side is zero if x and y are outside each other's light cone. First we take the continuum limit, $V \to \infty$:

$$\sum_{\vec{p}} \longrightarrow \int d_3p \, \frac{V}{(2\pi)^3}.$$

Calling the right hand side of the above commutator the function $\Delta_c(x-y)$ we have:

$$\Delta_c(x-y) = \frac{1}{(2\pi)^3} \int \frac{d_3p}{2p_0} \left\{ e^{ip(x-y)} - e^{-ip(x-y)} \right\}.$$

To prove that this is zero if the four vector $z = x - y$ is outside the light cone (which means $(zz) > 0$) we proceed in two steps. First we will show that Δ_c is Lorentz invariant. Then we will show that $\Delta_c(z)$ is zero for any z with $z_0 = 0$. Since all z outside the light cone can be obtained from some z with $z_0 = 0$ by a Lorentz transformation (that leaves Δ_c unchanged) we will have proven the required result.

The argument is illustrated in the figure. If the function is zero at the location of the cross it is zero in the whole shell going through that point, because all the points in the shell can be obtained from the cross point by means of a Lorentz transformation, and the function is Lorentz invariant. If the function is zero along the whole horizontal axis (the equal time line) then the function is zero everywhere outside the light cone.

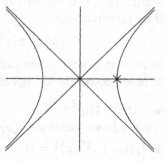

To show the Lorentz invariance we first rewrite the three-dimensional integral as a four-dimensional integral. This may be done as follows. Note that in the above $p_0 = \sqrt{\vec{p}^2 + m^2}$. We write

$$\frac{1}{2p_0} = \int_{-\infty}^{\infty} d\xi \, \delta\left(\xi^2 - \vec{p}^2 - m^2\right) \theta(\xi)$$

where

$$\theta(\xi) = \begin{cases} 1 & \text{if } \xi > 0 \\ 0 & \text{if } \xi < 0 \end{cases}$$

Indeed, the only points contributing to the integral are

$$\xi = \pm\sqrt{\vec{p}^2 + m^2}.$$

We may write

$$\delta\left(\xi^2 - \vec{p}^2 - m^2\right) = \delta\left\{\left(\xi - \sqrt{\vec{p}^2 + m^2}\right)\left(\xi + \sqrt{\vec{p}^2 + m^2}\right)\right\}$$

Now $\delta(ab) = \frac{1}{|a|}\delta(b) + \frac{1}{|b|}\delta(a)$. Furthermore the solution $\xi = -\sqrt{}$ will not give anything because the θ-function restricts us to positive values for ξ. We so get:

$$\int_{-\infty}^{\infty} d\xi \, \frac{1}{\xi + \sqrt{}} \delta(\xi - \sqrt{})\theta(\xi) = \frac{1}{2\sqrt{}}$$

which is indeed what we stated above.

Although it might cause some confusion we are going to use for ξ the name p_0, and the above result is

$$\frac{1}{2p_0} = \int dp_0 \, \delta\left(p_0^2 - \vec{p}^2 - m^2\right) \theta(p_0).$$

Using this the function Δ_c becomes:

$$\Delta_c(z) = \frac{1}{(2\pi)^3} \int d_4p \left\{e^{i(pz)} - e^{-i(pz)}\right\} \theta(p_0)\delta\left(p^2 + m^2\right)$$

where we used $\delta(a) = \delta(-a)$.

We now show that this integral is Lorentz invariant. Thus let $z = Lz'$, where L is some Lorentz transformation, and we will show that $\Delta_c(z) = \Delta_c(z')$. We have

$$\Delta_c(z) = \Delta_c(Lz')$$
$$= \frac{1}{(2\pi)^3} \int d_4p \left\{e^{i(pLz')} - e^{-i(pLz')}\right\} \theta(p_0)\delta\left(p^2 + m^2\right).$$

Now introduce four new variables $q_1 \ldots q_4$ related to the p by $p = Lq$. This is as if we did a Lorentz transformation on p, but it is really a change of integration variables. The integration volume element $d_4 p$ becomes $d_4 q$ times the Jacobian of the transformation:

$$d_4 p = det(L) d_4 q.$$

Since $det(L) = 1$ that gives no change. Furthermore:

$$(p, Lz') = (Lq, Lz') = (q, z')$$
$$p^2 = (p, p) = (Lq, Lq) = q^2.$$

Finally, what happens with $\theta(p_0)$? This is more subtle and requires a detailed investigation of the action of a Lorentz transformation on the vector p. This vector is restricted to values inside the upper light cone, because we must have $p^2 = -m^2$ (p is on mass-shell) and the θ-function restricts us to the upper light cone. Since any four-vector in the upper light cone transforms into another vector in the upper light cone also q will be in the upper light cone. Therefore also $\theta(q_0)$ will be non-zero if $\theta(p_0)$ was non-zero and zero if $\theta(p_0)$ was zero. In other words, $\theta(p_0) = \theta(q_0)$ if $p = Lq$, but for this the δ-function is crucial, because otherwise there would not be the restriction to the upper light cone.

The result of all this is:

$$\Delta_c(z) = \frac{1}{(2\pi)^3} \int d_4 q \left\{ e^{i(qz')} - e^{-i(qz')} \right\} \theta(q_0) \delta(q^2 + m^2) = \Delta_c(z').$$

The last step follows since this differs from the original expression only by a different notation for the integration variables.

Now step 2. Let us suppose $z_0 = 0$. Then

$$\Delta_c(z)_{z_0=0} = \frac{1}{(2\pi)^3} \int d_4 q \left\{ e^{i(\vec{q}\,\vec{z})} - e^{-i(\vec{q}\,\vec{z})} \right\} \theta(q_0) \delta(q^2 + m^2).$$

The $\theta\delta$ factor restricts the q_0 integration but not the $\vec{q}$ integration. The $\vec{q}$ integral goes from $-\infty$ to $+\infty$ for every component. Therefore

$$\int d_3 q \, f(\vec{q}) = \int d_3 q \, f(-\vec{q})$$

and we see that the two terms cancel and thus

$$\Delta_c(z)_{z_0=0} = 0.$$

This concludes our proof.

Another important property of the field ϕ is that it obeys the Klein–Gordon equation. The complete expression for the field $\phi(x)$ is:

$$\phi(x) = \sum_{\vec{p}} \frac{1}{\sqrt{2p_0 V}} \left\{ a(p)e^{i(px)} + a^\dagger(p)e^{-i(px)} \right\},$$

$$p_0 = \sqrt{\vec{p}^2 + m^2}.$$

It is easy to see that

$$\left(\Box - m^2 \right) \phi(x) = 0$$

due to the mass-shell relation for p_0. Finally, operating with $\phi(x)$ on the vacuum state we obtain the state for a particle located at the point $\vec{x}$ at time $x_0 = 0$:

$$\phi(x) \mid 0 > = \sum_{\vec{p}} \frac{1}{\sqrt{2p_0 V}} e^{i(px)} \mid p >.$$

Exercise 2.9 The commutation law derived above shows commutation of $\phi(x)$ and $\phi(y)$ for equal times:

$$[\phi(x), \phi(y)]_{x_0=y_0} = \Delta_c(z)_{z_0=0} = 0$$

with $z = x - y$. Show that

$$\left[\frac{\partial \phi(x)}{\partial x_4}, \phi(y) \right]_{x_0=y_0} = (\partial_4 \Delta_c)_{z_0=0} = -\delta_3 \left(\vec{x} - \vec{y} \right).$$

2.4 Structure of Hilbert Space

The connection between vectors in Hilbert space and measurements is in quantum mechanics through probability. Physical states correspond to vectors in Hilbert space of unit length. Length is defined by the usual dot-product in complex space. If a vector $\mid \xi >$ has the components $\xi_1, \xi_2 \ldots$, that is

$$\mid \xi > = \xi_1 \mid 1 > + \xi_2 \mid 2 > + \ldots$$

where $\mid 1 >, \mid 2 >$ etc. represent the first, second etc. basis vector then the dot-product of $\mid \xi >$ with itself is given by

$$\xi_1^* \xi_1 + \xi_2^* \xi_2 + \ldots.$$

This dot-product will be denoted as $< \xi \mid \xi >$. Similarly

$$< \eta \mid \xi > = \sum \eta_i^* \xi_i.$$

The connection with physical measurements is now as follows. Suppose we have a vector $| c >$ corresponding to some physical situation. Now let there be two other mutually orthogonal vectors $| a >$ and $| b >$, also corresponding to physical situations. Suppose now $| c >= \alpha | a > + \beta | b >$. Remember that $| a >, | b >$ and $| c >$ must all be of unit length, because they represent a physical state.

Now, if the system is in the state described by $| c >$ then, if one tries to measure whether the system is in state $| a >$ or $| b >$ one will find:

probability to find system in state $| a > = |\alpha|^2$

probability to find system in state $| b > = |\beta|^2$.

More generally, if a system is in a state $| c >$, then the probability to observe the state $| a >$ is given by

$$|< a \,|\, c >|^2 \,.$$

This is the fundamental connection between Hilbert space and physical measurements. Since $< c \,|\, c > = 1$ we have:

$$1 = \,< c \,|\, c > \,= \alpha^* \alpha < a \,|\, a > + \,\beta^* \beta < b \,|\, b > \,= \,|\,\alpha\,|^2 + |\,\beta\,|^2 \,.$$

The two probabilities add up to 1, as should of course be the case.

There is some potential trouble here in connection with Lorentz invariance, quite different from the trouble arising from dealing with a finite volume. As noted before, states in Hilbert space are generally not invariant under Lorentz transformations, but they transform according to some representation of these Lorentz transformations. The trouble now is that we certainly want probabilities to be invariant under Lorentz transformations, in other words, the above defined dot-product must be invariant. This dot-product involves complex conjugation. Now Lorentz transformations have i's in them, and so may be their representations. The Lorentz-invariant dot-product (such as $(pp) = \vec{p}^2 + p_4^2 = \vec{p}^2 - p_0^2$) involves no complex conjugation of that i (the essential point is here that the dot-product can be negative, not the particular use of i for this purpose). In order for the dot-product in Hilbert space to be Lorentz invariant we may have to deal with i's that should **not** be conjugated when forming dot-products. But this means that those dot-products are no more necessarily positive. We then must insist that states with negative probability are unphysical.

By itself this is not new: even in ordinary space there are restrictions, and momenta of particles are physical only if their dot-product with themselves is negative (corresponding to real mass). One has $(pp) = -m^2$, if m is the mass of the particle with momentum p. Particles with momenta such that (pp) is positive have not been seen. Some theorists have speculated on their existence, calling them tachyons, but the theory is probably not consistent with quantum mechanics.

Not conjugating the Lorentz i is our way for obtaining the minus sign in the dot-product and at the same time remaining manifestly (i.e., visible to the naked eye) Lorentz invariant. Since usually vectors appear only in dot-products where the fourth component of one vector is multiplied by the fourth component of another, that prescription is really nothing else but just a prescription, and it is not really so that the Lorentz i is not to be conjugated. The key point is that if k is some vector then k^* is not a vector, but $\bar{k}$ is (the bar indicates no complex conjugation of the Lorentz i of the fourth component). Thus if Lorentz invariance is to be manifest then one must use $\bar{k}$.

An important exception is when the 4-tensor $\epsilon_{\alpha\beta\mu\nu}$ occurs. This tensor, when contracted with four four-vectors, gives rise to an imaginary result, because precisely one of the indices will be 4. The rule is therefore that ϵ must be considered imaginary. All together, our convention is that we will not conjugate the Lorentz i when occurring in a vector, but the ϵ tensor is taken to be imaginary.

So, much the same as in ordinary space were some domains seem to be excluded for physics, the same might happen in Hilbert space, where we would be truly in difficulty if we had to allow for physical systems with negative probability. No consistent theory can then be constructed, because probability is positive by its very definition.

For scalar or pseudoscalar particles there is not yet any problem, the tranformations in Hilbert space are rather trivial (as for example the Hilbert space vector $|\vec{p}>$ transforming to the vector $|\vec{q}>$ where $\vec{q}$ is the Lorentz transform of $\vec{p}$). In dealing with particles with spin, such as electrons and photons, the transformation of the states in Hilbert space becomes more complicated, and we will actually run into this problem.

3
Interacting Fields

3.1 Physical System

We will now discuss systems of interacting particles. The formalism is the Lehmann–Symanzik–Zimmermann formalism, and we are using Källén's method to derive the S-matrix from the equations of motion.

For definiteness assume the existence of two kinds of particles called π and σ. Both are spinless, that complication is for later. We assume masses M and m for the π and σ particles respectively. The whole is a simplified version of electrodynamics of electrons and photons. The σ will play the role of the photon.

To begin we will focus on a specific problem, namely $\pi\pi$ scattering. The physical process is as in the figure.

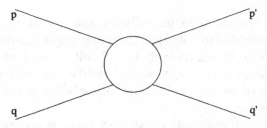

Two pions with momenta p and q (with $p^2 = q^2 = -M^2$) meet and scatter, and we are interested in the probability that a final configuration of two pions with momenta p' and q' (with $p'^2 = q'^2 = -M^2$) is produced. This probability, when multiplied with the appropriate flux factors, is the differential cross section for this process.

3.2 Hilbert Space

This needs very careful consideration. A vector in Hilbert space represents a physical state. What is a physical state? A physical state is simply a possible physical situation, with particles moving here and there, with collisions, with dogs chasing cats, with people living and dying, with all kinds of things happening. Often people make the mistake of identifying a physical state with the situation at a given moment, one picture from a movie, but that is not what we call the physical state. The situation at some moment may be seen as a boundary condition: if one knows the whole situation at some moment, and one knows the laws of nature, then in principle we can deduce the rest.

Thus a physical state may be **characterized** by the situation at a definite moment, but the state itself refers to the **whole world including the progress in time**. Conveniently, especially for scattering processes one may use the time points $\pm\infty$. Thus the above process corresponds to a vector in Hilbert space, and we can denote that vector by

$$\mid p, q >_{\text{in}} .$$

By this we mean: that physical system that has two pions of momenta p and q at time $t = -\infty$ (the "in" configuration). It must be understood that $\mid p, q >_{\text{in}}$ contains everything, including how the system looks at other times. For example, we could define the state $\mid p, q >_0$ as that physical system which has at time $t = 0$ exactly two pions of momenta p and q. The above described state, $\mid p, q >_{\text{in}}$, has two such pions at $t = -\infty$, but it may well be that they scatter before $t = 0$, and the probability that we have still two pions with that same momentum at $t = 0$ is smaller than one.

The above discussion leads up to the following identification. Let $\mid p, q >_{\text{in}}$ and $\mid p, q >_0$ be the systems as described above. They are different systems. Let the system be in the state $\mid p, q >_{\text{in}}$ (two pions at $t = -\infty$). The probability of having two pions with momenta p and q at $t = 0$ is the square of the absolute value of the dot-product between the states:

$$\mid _0< p, q \mid p, q >_{\text{in}} \mid^2.$$

If some collision took place before $t = 0$ we may actually still have two pions but with different momenta, say k and r. A state with two pions with momenta k and r at time $t = 0$ is denoted by $\mid k, r >_0$. If the system is in the state $\mid p, q >_{\text{in}}$ then the probability of observing two pions of momenta k and r at time 0 is given by:

$$\mid {}_0< k, r \mid p, q >_{\text{in}} \mid^2.$$

Thus the state $\mid p, q >_{\text{in}}$, when viewed at time $t = -\infty$ contains two pions of momenta p and q, but if we look to it at time $t = 0$ we see with some probability two pions that may or may not have the momenta p and q. More generally, new particles may be produced in a collision, so we may also see three, four, etc. pion configurations. For example,

$$\mid {}_0< k, r, s \mid p, q >_{\text{in}} \mid^2$$

gives the probability of observing three pions of momenta k, r and s at time $t = 0$ if we know that the system is in a state characterized by the fact that at time $t = -\infty$ there were two pions of momenta p and q.

3.3 Magnitude of Hilbert Space

It follows from the above that we can have as many states in Hilbert space as possible situations at $t = -\infty$. We now **assume** the following asymptotic condition: if particles are sufficiently far apart in space they do not interact and behave as free particles. If we go sufficiently far back in time particles will be separated. Therefore we may assume that the possible situations at $t = -\infty$ are precisely the situations of non-interacting particles. Thus the Hilbert space of interacting systems is by this assumption equally large as the Hilbert space of free particles.

Clearly we will have to modify this if we want to consider stable bound states. No matter how far back we go in time, the electron and proton in a hydrogen atom do not separate. To describe such situations properly we must enlarge Hilbert space and allow states containing hydrogen atoms. Of course, such atoms again can just be considered as a new kind of particle, and the Hilbert space becomes then effectively the free Hilbert space of three (in this case) particles (electrons, protons and hydrogen atoms). We will concentrate however on simple particle states, although at some point we should investigate how these bound states fit in the picture.

Now that we are clear about the meaning of states and their representation in Hilbert space we can proceed and postulate equations that will describe particles in interactions. Experiment must then decide which equations describe nature. Of course, whatever we postulate, it will be within the framework of Lorentz invariant quantum mechanics. Only a limited degree of freedom is left.

3.4 *U*-matrix, *S*-matrix

In describing scattering problems, where one typically considers initial and final configurations of widely separated particles, the description in terms of "in" and "out" states is advantageous. A vector $| a >_\text{in}$ in Hilbert space corresponds to a physical system characterized by the configuration a at time $t = -\infty$. Similarly a vector $| b >_\text{out}$ corresponds to a system characterized by the configuration b at time $t = +\infty$. The states $| a >_\text{in}$ can be counted in the same way as the free particle states:

$$| 0 >_\text{in} \quad \text{vacuum at } t = -\infty$$

$$| \vec{p}_1 >_\text{in} \quad \text{one particle (pion) at } t = -\infty$$

$$\vdots$$

$$| \vec{p}_1, \vec{p}_2, \ldots \vec{q}_1, \vec{q}_2 \ldots >_\text{in} \quad \text{pions with momenta } \vec{p}_1, \ \vec{p}_2, \ldots \text{ and}$$
$$\sigma\text{-particles with momenta } \vec{q}_1, \vec{q}_2, \ldots$$
$$\text{at time } t = -\infty.$$

A remark needs to be made here. Since for free particles the energy is known if the three-momentum is known the state is characterized by the three-momenta only. That is why we used the three-vector as argument in $| \vec{p}_1 >_\text{in}$ rather than $| p_1 >_\text{in}$. In the following we will often drop the arrow, assuming that the reader is aware that the particles indicated are on mass shell. Note that for finite times, when the particles are not necessarily far apart, the energy is not simply given by the usual mass shell relation.

Similarly "out" states. Of course, in general if there is interaction the "in" states are different from the "out" states, although both are in the same Hilbert space. Thus

$$| \vec{p}_1, \vec{p}_2 >_{\text{in}} \neq | \vec{p}_1, \vec{p}_2 >_{\text{out}}$$

because a system that has at $t = -\infty$ two pions with momenta $\vec{p}_1$ and $\vec{p}_2$ is unlikely to still have two pions with momenta $\vec{p}_1$ and $\vec{p}_2$ at $t = +\infty$. There is some probability that the pions do not scatter; it is given by the absolute value squared of the dot product between the two states:

$$| _{\text{out}}< \vec{p}_1, \vec{p}_2 | \vec{p}_1, \vec{p}_2 >_{\text{in}} |^2$$

is the probability that, starting with two pions with momenta $\vec{p}_1, \vec{p}_2$ at $t = -\infty$ we will still find two pions with momenta $\vec{p}_1$ and $\vec{p}_2$ at $t = +\infty$. Similarly

$$| _{\text{out}}< \vec{p}', \vec{q}' | \vec{p}, \vec{q} >_{\text{in}} |^2$$

is the probability that when measuring on a system characterized by there being two particles of momenta $\vec{p}$ and $\vec{q}$ at time $t = -\infty$ one will find two particles of momenta $\vec{p}'$ and $\vec{q}'$ at $t = +\infty$.

We thus have two sets of basis vectors in the same Hilbert space, namely the "in" basis ($| 0 >_{\text{in}}, | \vec{p}_1 >_{\text{in}} \dots$) and the "out" basis ($| 0 >_{\text{out}}, | \vec{p}_1 >_{\text{out}} \dots$). Since a system without any particles at $t = -\infty$ will not have any particles at $t = +\infty$ we have $| 0 >_{\text{in}} = | 0 >_{\text{out}}$. Similarly for one particle states, $| \vec{p} >_{\text{in}} = | \vec{p} >_{\text{out}}$. But for two or more particle states this is not true if there is interaction.

Since physical states correspond to vectors of unit length both the "in" and "out" bases are orthonormal. There will be a matrix that transforms the "in" basis into the "out" basis:

$$| a >_{\text{out}} = S^\dagger | a >_{\text{in}},$$

for any configuration a. We have used $S^\dagger$ (complex conjugation and reflection) here purely by convention. The S-matrix contains all physical information for any scattering process. For instance, the probability to have configuration a at $t = -\infty$ and to find b at $t = +\infty$ is given by

$$| _{\text{out}}< b | a >_{\text{in}} |^2 = | _{\text{in}}< a | b >_{\text{out}} |^2 = | _{\text{in}}< a | S^\dagger | b >_{\text{in}} |^2$$
$$= | _{\text{in}}< b | S | a >_{\text{in}} |^2.$$

The first step is because $< a | b > = < b | a >^*$, and the last step is nothing but the very definition of $S^\dagger$ relative to S. Since both "in" and "out" states are orthonormal sets the S-matrix is unitary (dot-products are unchanged):

$$S^\dagger S = 1.$$

Later we will interpret this as conservation of probability. If there is no interaction the S-matrix is 1. One therefore often writes

$$S = 1 + iT$$

and the relation $S^\dagger S = 1$ becomes

$$i\left(T - T^\dagger\right) = -T^\dagger T$$

Exercise 3.1 Verify this.

Exactly like for free particle states we may define matrices a and $a^\dagger$ in Hilbert space. We can do that on both "in" and "out" bases. Thus, $a_{in}(\vec{p})$ is a matrix such that

$$a_{in}(\vec{p}) \mid \vec{p}, \vec{p} \ldots \vec{q}, \ldots >_{in} = \sqrt{n} \mid \vec{p}, \vec{p} \ldots \vec{q}, \ldots >_{in}$$
$$n \text{ times } \vec{p} \qquad n - 1 \text{ times } \vec{p}.$$

Similarly $a_{out}(\vec{p})$ is defined by its action on the unit vectors of the "out" basis. Note that at this point we have no idea what happens when we apply $a_{in}(\vec{p})$ on some "out" basis vector.

From the $a(\vec{p})$ we may construct fields $\phi_{in}(x)$ and $\phi_{out}(x)$:

$$\phi_{in}(x) = \sum \frac{1}{\sqrt{2Vp_0}} \left\{ a_{in}(\vec{p})e^{ipx} + a_{in}^\dagger(\vec{p})e^{-ipx} \right\}.$$

The momentum p refers here to a momentum that can be found with the particles at $t = -\infty$, when they are presumably far apart, and thus their energy will be purely kinetic, no potential energy. Thus then $p_0 = \sqrt{\vec{p}^2 + M^2}$ for pions (and $p_0 = \sqrt{\vec{p}^2 + m^2}$ for σ-particles, with which we have the field $\sigma_{in}(x)$).

Since the S-matrix transforms "in" into "out" basis it also transforms a_{in} into a_{out}:

$$S\, a_{out}(\vec{p})S^\dagger \mid \vec{p} \ldots >_{in}$$
$$= S\, a_{out}(\vec{p}) \mid \vec{p} \ldots >_{out} = \sqrt{n}\, S \mid \vec{p} \ldots >_{out}$$
$$= \sqrt{n} \mid \vec{p} \ldots >_{in}.$$

We used here $S^{-1} = S^\dagger$, thus transforming "out" into "in" state. We see that

$$a_{in}(\vec{p}) = S\, a_{out}(\vec{p})\, S^\dagger$$

Similarly

$$\phi_{\text{in}}(x) = S\,\phi_{\text{out}}(x)\,S^\dagger.$$

Both fields, ϕ_{in} and ϕ_{out}, satisfy the Klein–Gordon equation

$$\left(\Box - M^2\right)\phi_{\text{in}}(x) = 0, \qquad \left(\Box - M^2\right)\phi_{\text{out}}(x) = 0.$$

It must be understood that ϕ_{in} and ϕ_{out} are well defined for all space and time. Thus $\phi_{\text{in}}(x)$ for example is perfectly well defined and non-zero for $x_0 = +\infty$.

The essence of the above may be put as follows: The assumption that for $t = -\infty$ any physical system becomes a system of free particles allows a mapping of all possible physical systems (by how they are at $-\infty$) on all possible free particle systems. Then we can use the formalism developed for those systems, and build fields. The fields so constructed are the "in" fields. Similarly "out" fields, related to labeling physical systems by how they are at $+\infty$. This then exhibits the role of assumptions on asymptotic behaviour, clearly of fundamental importance.

Let us formulate this once more in different terms. Consider a physical system as a movie, showing a lot of action as it is projected. This movie is in a can, and the can may be labelled in different ways. The labelling may be on the basis of the opening scene of the movie, or alternatively, on the basis of the final scene. Clearly, these are very different ways of labelling. Only empty movies (the vacuum) or movies containing just one actor (one particle of some momentum) are likely to have identical "in" and "out" scenes.

A physical state in Hilbert space is like a movie in a can. It is the whole movie, not just the opening scene, even if the can is labelled that way. Seeing things this way it becomes hopefully clear that a progressing physical situation is not a vector in Hilbert space (such as $|\,\vec{p}\vec{q}>_{\text{in}}$) rotating to another state ($|\,\vec{p}\vec{q}>_{\text{out}}$) in the course of time. A vector in Hilbert space has no time dependence, but, like in a movie, all action is contained in that state. In that sense the S-matrix is a cross-index register, showing the relation between two labelling systems. Given the beginning scene of a movie, the S-matrix tells us what the final scene is.

3.5 Interpolating Fields

We are now in a position to formulate equations of motion that describe an interaction. We assume the existence of a field $\phi(x)$ (no "in" or "out" index) that is equal to $\phi_{in}(x)$ if x_0 (the time) is $-\infty$, and is equal to ϕ_{out} at $x_0 = +\infty$. Again $\phi_{in}(x)$ is well defined at $x_0 = +\infty$, but then it will be very different from $\phi(x)$. Thus

$$\lim_{x_0 \to -\infty} \phi(x) = \phi_{in}(x), \qquad \lim_{x_0 \to +\infty} \phi(x) = \phi_{out}(x).$$

The figure shows some attempt to visualize the situation. Note that ϕ_{in} and ϕ_{out} are well defined for all times. Now also $\phi(x)$ will satisfy an equation of motion, but it will not be a simple Klein–Gordon equation. We write:

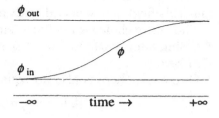

$$\left(\Box - M^2 \right) \phi(x) = -j(x).$$

The minus sign is to follow usual conventions. The quantity $j(x)$ is called a **current**, and $j(x)$ is such that if we solve this equation with the boundary condition $\phi(x) = \phi_{in}(x)$ at $x_0 = -\infty$, then for $x_0 = +\infty$ we will have $\phi(x) = \phi_{out}(x)$. If we know $j(x)$, and can solve this equation then we can find ϕ_{out} from ϕ_{in}, and thus also determine the S-matrix since S relates in- and out-fields.

Unfortunately the equation can be solved only in successive approximations. We start by assuming that $j(x)$ can be constructed from the ϕ themselves. This assumption has its basis in the fact that the a and $a^\dagger$ are matrices that allow construction of any other matrix, while $\phi(x)$ is that combination that is also local. Thus if we want the time development of $\phi(x)$ to be local (nothing moving with speeds exceeding the speed of light) then it makes sense to build up $j(x)$ from the $\phi(x)$.

Basically all this is one big assumption. The system is really very complicated, even for the simplest cases as we will see. Only for those simple cases can the above equation be solved, and even then only in terms of successive iterations (perturbation theory). Thus the scheme developed below is to a large extent determined simply by the requirement that we can solve it. Fortunately these methods give rise to results that agree very well with experimental observations. One truly may be thankful to nature for this: limiting itself to something that we can compute!

Let us now write down some simple expression for $j(x)$ and solve the equation. Since we want the pions to interact with the σ-particles we will also include σ-fields in $j(x)$. We **assume**

$$j(x) = 2g\sigma(x)\phi(x).$$

The constant g is called the coupling constant. In the following we will write $\pi(x)$ instead of $\phi(x)$ to exhibit more clearly the fact that this field is associated with pions *.

The above choice is really the simplest non-trivial form for $j(x)$, choosing simply $\phi(x)$ or $\sigma(x)$ one can by some reshuffling make $j(x)$ zero.

Of course, there is also an equation of motion for the σ-field:

$$\left(\Box - m^2\right)\sigma(x) = -\bar{j}(x)$$

with some other current $\bar{j}(x)$ instead of $j(x)$. As we will see one cannot freely choose $\bar{j}$ if j is already fixed. There is an interconnection. Intuitively this is simple to see: if $j(x)$ determines how pions interact with σ's then evidently this fixes also the interaction of σ's with π's. In fact, one can determine the S-matrix from either j or $\bar{j}$ above, and this better be the same S-matrix! We will see that the choice $j = 2g\sigma\pi$ implies $\bar{j} = g\pi^2$.

It should be stressed here that the field $\sigma(x)$ in j is not σ_{in} or σ_{out} but also the "interpolating" field (interpolating between σ_{in} and σ_{out}) just as $\pi(x)$ is the interpolating field (interpolating between π_{in} and π_{out}).

To find the S-matrix we assume the existence of a matrix $U(x_0)$, time but not space dependent, and such that:

$$\pi(x) = U^\dagger(x_0)\pi_{\text{in}}(x)U(x_0)$$

$$\sigma(x) = U^\dagger(x_0)\sigma_{\text{in}}(x)U(x_0).$$

Clearly $U(-\infty) = 1$, and $U(+\infty) = S$. Since we will find $U(x_0)$ there is no point in discussing whether this assumption makes sense. To be complete, it should be noted that from a strict

---

* Sometimes in books on field theory the notation $\pi(x) = i\partial_4\phi(x)$ is used. That has nothing to do with our $\pi(x)$.

mathematical point of view one can raise many questions and objections, relating to the fact that we are dealing with infinite matrices.

From the fact that $\pi(x)$ satisfies an equation of motion we should be able to deduce an equation of motion for $U(x_0)$. Here we need to be careful because all the objects that we are dealing with are big, generally non-commuting, matrices. A basic equation, derived before, that we will use is:

$$[\partial_4\pi(x), \pi(y)]_{x_0=y_0} = -\delta_3(\vec{x} - \vec{y}).$$

We will now show that, if $U(x_0)$ satisfies the following differential equation:

$$\frac{\partial U(x_0)}{\partial x_4} = g \int d_3y \; \pi_{\text{in}}^2(y)\sigma_{\text{in}}(y)U(x_0), \quad \text{with } y_0 = x_0,$$

then $\pi(x)$ as defined above satisfies

$$\left(\Box - M^2\right)\pi(x) = -2g\pi(x)\sigma(x).$$

The proof of this statement is not particularly difficult, just a little cumbersome. It is important to remember that $U(x_0)$ is time dependent, not space dependent. In other words

$$\partial_\mu U(x_0) = 0 \text{ for } \mu = 1, 2, 3.$$

First we introduce a notation:

$$H(x_0) = g \int d_3y \; \pi^2(y)\sigma(y), \quad \text{with } y_0 = x_0,$$

and

$$H_{\text{in}}(x_0) = g \int d_3y \; \pi_{\text{in}}^2(y)\sigma_{\text{in}}(y), \quad \text{with } y_0 = x_0.$$

It follows that

$$H(x_0) = U^{-1}H_{\text{in}}(x_0)U.$$

The equation to be solved is now:

$$\partial_4 U(x_0) = H_{\text{in}}(x_0)U(x_0).$$

The time derivative of $\pi(x)$ and $\sigma(x)$ can be computed. Remember:

$$0 = \partial(1) = \partial\left(U^{-1}U\right) = \left(\partial U^{-1}\right)U + U^{-1}\partial U$$

or

$$\partial U^{-1} = -U^{-1}(\partial U)U^{-1}.$$

We find:
$$\partial_4 \pi = -U^{-1} \partial_4 U U^{-1} \pi_{\text{in}} U + U^{-1} \partial_4 \pi_{\text{in}} U + U^{-1} \pi_{\text{in}} \partial_4 U$$
$$= -U^{-1} H_{\text{in}} \pi_{\text{in}} U + U^{-1} \partial_4 \pi_{\text{in}} U + U^{-1} \pi_{\text{in}} H_{\text{in}} U$$
$$= U^{-1} [\pi_{\text{in}}, H_{\text{in}}] U + U^{-1} \partial_4 \pi_{\text{in}} U.$$

We see a general rule here. We need the second time derivative of the π-field:
$$\partial_4^2 \pi = U^{-1} [[\pi_{\text{in}}, H_{\text{in}}], H_{\text{in}}] U + U^{-1} \partial_4 [\pi_{\text{in}}, H_{\text{in}}] U$$
$$+ U^{-1} [\partial_4 \pi_{\text{in}}, H_{\text{in}}] U + U^{-1} \partial_4^2 \pi_{\text{in}} U.$$

As noted before the spatial derivatives on U vanish. Considering now the Klein–Gordon equation for $\pi(x)$ we find:
$$\left(\Box - M^2 \right) \pi(x) = \left(\Box - M^2 \right) U^{-1} \pi_{\text{in}} U$$
$$= U^{-1} \left\{ \left(\Box - M^2 \right) \pi_{\text{in}} \right\} U$$
$$+ U^{-1} [[\pi_{\text{in}}, H_{\text{in}}], H_{\text{in}}] U + U^{-1} \partial_4 [\pi_{\text{in}}, H_{\text{in}}] U$$
$$+ U^{-1} [\partial_4 \pi_{\text{in}}, H_{\text{in}}] U.$$

Since the field π_{in} satisfies the free Klein–Gordon equation, the first term on the right hand side vanishes. Also the second and third terms vanish because $\pi_{\text{in}}(x)$ and $\pi_{\text{in}}(y)$ commute for $x_0 = y_0$ (as far as $\pi_{\text{in}}(x)$ and $\sigma_{\text{in}}(y)$ are concerned, they commute always because the matrices a and $a^\dagger$ for the π and σ fields commute). The last term gives
$$[\partial_4 \pi_{\text{in}}, H_{\text{in}}] = g \int d_3 y \, \sigma(y) \left[\partial_4 \pi_{\text{in}}(x), \pi_{\text{in}}^2(y) \right]_{x_0 = y_0}$$
$$= -2g \int_{x_0 = y_0} d_3 y \, \sigma_{\text{in}}(y) \delta_3 (\vec{x} - \vec{y}) \pi_{\text{in}}(y) = -2g \sigma_{\text{in}}(x) \pi_{\text{in}}(x).$$

The final result is:
$$\left(\Box - M^2 \right) \pi(x) = -2g U^{-1} \sigma_{\text{in}}(x) \pi_{\text{in}}(x) U = -2g \sigma(x) \pi(x)$$

because
$$U^{-1} \sigma_{\text{in}} \pi_{\text{in}} U = U^{-1} \sigma_{\text{in}} U U^{-1} \pi_{\text{in}} U = \sigma \pi$$

which is precisely what we set out to prove.

Exercise 3.2 Verify the equation
$$\left(\Box - m^2 \right) \sigma(x) = -g \pi^2(x).$$

This is the moment to consider the question of the connection between j and $\bar{j}$. It is clear from the above derivation that they follow from the same U, i.e., from the same H. As a matter of fact one notes the formal rules:

$$j(x) = \frac{\partial}{\partial\pi}\mathcal{H}(x) \quad \text{and} \quad \bar{j}(x) = \frac{\partial}{\partial\sigma}\mathcal{H}(x)$$

with

$$H = \int d_3y\ \mathcal{H}(y), \qquad \mathcal{H}(y) = g\pi^2(y)\sigma(y).$$

The quantities H and $\mathcal{H}$ are called the **interaction Hamiltonian** and the **interaction Hamiltonian density** respectively. It is clear that it is better to start from a Hamiltonian and then to derive the equations of motion to avoid inconsistencies. This is what we will do in general. In fact, at this point we do not really need the equations of motion for the fields any more. We will simply take some Hamiltonian, and we then know that the U-matrix satisfies the equation

$$\partial_4 U(x_0) = H_{\text{in}}(x_0)U(x_0)$$

and solve U from that. Once we have U we have the S-matrix, namely $S = U(\infty)$.

Incidentally, we will use the words Hamiltonian and Lagrangian interchangeably. The interaction Hamiltonian and Lagrangian are not the same, they are different if there are derivatives, but there is no need to entangle ourselves in these outdated questions. The interaction Hamiltonian is the quantity that appears in the U-matrix, but eventually we will abandon the U-matrix and use the interaction Lagrangian to specify the interactions of the theory.

It should be clearly understood what we have here. The solution of the equation for U will give us the matrix U as a function of π_{in} and σ_{in}. Thus we will obtain S as a function of the π_{in} and σ_{in}. This is precisely what we need. As noted before, the probability to find the configuration c at time $t = \infty$ if at time $t = -\infty$ we have the configuration b, is given by

$$|\ _{\text{out}}\!<c\mid b>_{\text{in}}|^2 = |\ _{\text{in}}\!<c\mid S\mid b>_{\text{in}}|^2.$$

We need $_{\text{in}}\!<c\mid S\mid b>_{\text{in}}$. If S is given in terms of "in" fields we know exactly what S gives when operating on an "in" basis vector.

In principle there is thus no problem here. However, the actual calculation of matrix elements $_{in}< c \mid S \mid b >_{in}$ remains a complicated matter. The diagram technique introduced by Stückelberg and Feynman is a very powerful tool to analyze this situation. This technique is generally useful for solving equations of the type

$$\frac{\partial}{\partial t}X = AX$$

where A is a given, time dependent, quantity. If A is a constant the solution is simple:

$$X = e^{At}.$$

If A is time dependent, but not a matrix, just some function, then the solution is:

$$X = \exp\left(\int_c^t A(t')dt'\right)$$

where c is an arbitrary constant. If $A(t)$ is a matrix such that $A(t_1)$ and $A(t_2)$ do not necessarily commute then the solution is more complicated and essentially can be given only in terms of a series expansion that looks very much like an expansion of the above exponential, but not exactly. To be precise, one finds

$$X = 1 + \sum_{n=1}^{\infty} \frac{1}{n!} \int_{-\infty}^{t} dt_1 \int_{-\infty}^{t} dt_2 \ldots \int_{-\infty}^{t} dt_n T\left(A(t_1)A(t_2)\ldots\right).$$

We have taken the constant c as $-\infty$, to avoid a number of irrelevant complications. The "time ordered" product T is defined as follows. For the product of two A's:

$$T\left(A(t_1)A(t_2)\right) = \begin{cases} A(t_1)A(t_2) & \text{if } t_1 > t_2 \\ A(t_2)A(t_1) & \text{if } t_1 < t_2. \end{cases}$$

For any number of A's similarly: the A must be arranged in order of decreasing time.

Obviously, if $A(t_1)$ and $A(t_2)$ commute for any t_1 and t_2 then the above expression reduces to an exponential.

We now proceed to show the correctness of the above solution. It can of course be verified directly by putting the above solution into the equation, but we will also give the standard derivation, which is by iteration.

Exercise 3.3 Verify directly the given solution.

Suppose we want to find X as a power series in A. Let us find the lowest order term. We write $X = 1 + \alpha_1$, where α_1 is of first order in A. Neglecting terms of order A^2 such as $A\alpha_1$ the equation for X becomes

$$\frac{\partial \alpha_1}{\partial t} = A(t), \quad \text{which gives} \quad \alpha_1 = \int_{-\infty}^{t} dt_1 A(t_1).$$

To obtain the next iteration we write:

$$X = 1 + \int_{-\infty}^{t} dt_1 A(t_1) + \alpha_2$$

where α_2 is of second order in A. The equation for X becomes (to second order in A):

$$\frac{d\alpha_2}{dt} = A(t) \int_{-\infty}^{t} dt_1 A(t_1).$$

The solution is

$$\alpha_2 = \int_{-\infty}^{t} dt_2 A(t_2) \int_{-\infty}^{t_2} dt_1 A(t_1).$$

Notice that $t_2 > t_1$, the matrices A appear in descending order of time. To further proceed with this integral we claim that also

$$\alpha_2 = \int_{-\infty}^{t} dt_2 \int_{t_2}^{t} dt_1 A(t_1) A(t_2).$$

Note the order of the A and the integration limits of the second integral where again the A appear in descending order. This may be verified either by direct insertion into the equation for $d\alpha/dt$ or by transforming the integral. This becomes very easy by considering a figure showing the integration domains. Taking c as lower integration limit, with c some number, the first integral 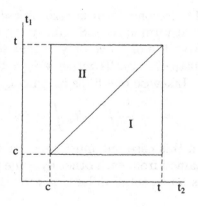 corresponds to domain I in the figure and the second to domain II. It is clear that the two domains can be obtained from each other by exchanging t_1 and t_2. Since indeed the integrands have t_1 and t_2 interchanged we see that the two integrals are equal. We may therefore take also for α_2 half the sum of both expressions and thus obtain

$$\alpha_2 = \tfrac{1}{2} \int_{-\infty}^{t} dt_1 \int_{-\infty}^{t} dt_2 \, T\left(A(t_1)A(t_2)\right).$$

For the sake of clarity we will show directly that the above is the correct solution. First,

$$T\left(A(t_1)A(t_2)\right) = \theta(t_1 - t_2)A(t_1)A(t_2) + \theta(t_2 - t_1)A(t_2)A(t_1)$$

where $\theta(x) = 1$ if $x > 0$ and $= 0$ if $x < 0$. This shows explicitly the meaning of the T-product. Now,

$$\frac{d}{dt}\alpha_2 = \tfrac{1}{2} \int_{-\infty}^{t} dt_2 \left[\theta(t - t_2)A(t)A(t_2) + \theta(t_2 - t)A(t_2)A(t)\right]$$

$$+ \tfrac{1}{2} \int_{-\infty}^{t} dt_1 \left[\theta(t_1 - t)A(t_1)A(t) + \theta(t - t_1)A(t)A(t_1)\right].$$

Essentially the first part is obtained by setting $t_1 = t$, the second by setting $t_2 = t$. The very first term contains $\theta(t - t_2)$ and is zero unless $t_2 < t$ which is always true since t_2 runs from $-\infty$ to t. We can therefore omit the θ-function in that term, to get

$$\tfrac{1}{2} A(t) \int_{-\infty}^{t} A(t_2) dt_2.$$

The second term is zero unless $t_2 > t$ which is never true, and that term is zero. Similarly the last two terms, which really differ from the first two only in that the integration variable is called t_1 instead of t_2. Together we get the desired result.

Likewise one finds in general:

$$\alpha_n = \frac{1}{n!} \int_{-\infty}^{t} dt_1 \int_{-\infty}^{t} dt_2 \ldots \int_{-\infty}^{t} dt_n \, T\left(A(t_1)A(t_2)\ldots\right).$$

In this case one must consider $n!$ domains of integration, all obtained from each other by some permutation of $t_1 \ldots t_n$, but there is no essential difference with the case of two variables.

3.6 Feynman Rules

We will now work out the lowest non-vanishing order of the S-matrix for the case of the π and σ fields given before. We also must be careful with the i from $x_4 = ix_0$. We have

$$\partial_4 U = H_{\text{in}} U$$

or

$$\partial_0 U(x_0) = i \int_{y_0 = x_0} d_3 y \, \mathcal{H}_{\text{in}}(y) \, U(x_0).$$

In first approximation we find:

$$U(x_0) = 1 + i \int_{-\infty}^{x_0} dt_1 \int_{y_0 = t_1} d_3 y \, \mathcal{H}_{\text{in}}(y)$$

where we used y_0 directly instead of t_1. Note that $d_4 y = d_3 y \, dy_0$. The S-matrix follows by taking the limit $x_0 = +\infty$:

$$S = 1 + i \int d_4 y \, \mathcal{H}_{\text{in}}(y) + \dots$$

The general expression for S is:

$$S = 1 + \sum_{n=1}^{\infty} \frac{i^n}{n!} \int d_4 y_1 d_4 y_2 \dots d_4 y_n \, T\left(\mathcal{H}_{\text{in}}(y_1) \dots \mathcal{H}_{\text{in}}(y_n)\right).$$

Let us now concentrate on a specific process. Consider the scattering of two pions with momenta p and q giving rise to two pions with momenta p' and q'. We must calculate:

$$_{\text{out}}< p', q' \mid pq >_{\text{in}} \; = \; _{\text{in}}< p', q' \mid S \mid pq >_{\text{in}}$$

with S as above, and

$$\mathcal{H}_{\text{in}}(y) = g\pi_{\text{in}}^2(y)\sigma_{\text{in}}(y).$$

Remember:

$$\pi_{\text{in}}(y) = \sum \frac{1}{\sqrt{2Vk_0}} \left\{ a_{\text{in}}(k)e^{iky} + a_{\text{in}}^\dagger(k)e^{-iky} \right\}$$

where $a_{\text{in}}(k)$ transforms a state with m pions into a state with $m - 1$ pions, and gives zero if no pions are present, while $a_{\text{in}}^\dagger(k)$ gives the opposite.

Let us first consider the lowest order term of S. As we have now exclusively "in" type objects we will drop this subscript. We have:

$$< p'q' \mid S \mid pq > \; = \; < p'q' \mid pq > + i \int d_4 y < p'q' \mid \mathcal{H}(y) \mid pq >$$

plus terms of higher order in $\mathcal{H}$. Now, if p', q' is different from p, q then the first term is zero (orthogonal vectors). The second term contains one $\mathcal{H}$ and therefore only one σ field. This applied to a state without σ particles gives zero (for the $a(k)$ part) or a state containing a σ particle. But the dot product of such a state with the state $\mid p'q' >$, containing no σ particle is zero. Therefore also the second term is zero. Generally, any product of an odd number of $\mathcal{H}$ gives zero between states without σ-particles, by similar arguments.

Let us now consider the second order term. It is given by:

$$\frac{i^2 g^2}{2} \int d_4 y \, d_4 y' < p'q' \mid T \left\{ \pi^2(y)\sigma(y)\pi^2(y')\sigma(y') \right\} \mid pq > .$$

Now we have non-zero terms. For instance, some $a^\dagger$ term in the σ-field can transform the state $\mid pq >$ into a state containing in addition a σ particle of, say, momentum k:

$$\sigma \mid pq > = \mid pq, k > .$$

Then two a-type terms in the following two π fields may transform this state into a state containing no pions:

$$\pi^2 \sigma \mid pq > = \mid k > .$$

Next the appropriate $a(k)$ term in the σ field transforms this state into the state $\mid 0 >$

$$\sigma \pi^2 \sigma \mid pq > = \mid 0 > .$$

Finally, selecting the terms with $a^\dagger(p')$ and $a^\dagger(q')$ in the last two pion fields transforms the state $\mid 0 >$ into the state $\mid p'q' >$. The dot product of this state with $\mid p'q' >$ is non-zero, in fact is one.

This may be graphically depicted in the following way. Particles are described by lines, and the action of σ and π fields is to either end or start a line. The action of $\mathcal{H}$ is thus to start or end two π and one σ line. The above example, drawn in the opposite direction (i.e. with y' left of y) is:

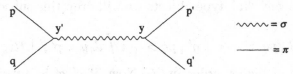

$\mathcal{H}(y')$ ends two π lines and starts a σ line corresponding to $\pi^2 \sigma \mid pq > = \mid k >$ and $\mathcal{H}(y)$ ends a σ line and starts two π lines. We now can draw pictures corresponding to all possibilities.

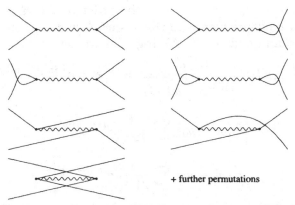

+ further permutations

Exceptionally, we have drawn the vertices as visible dots, to avoid confusion. The last case shown differs from the first only by the interchange of y and y'. Since the whole is symmetric in y and y' it follows that both cases give the same. Also the first four diagrams all correspond to the same expression. All together we get:

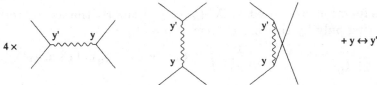

Generally, in higher orders one gets the same for all permutations of $y, y', y'', \ldots$, which gives a factor $n!$ for the nth order. This cancels against the factor $1/n!$ in front.

So far we have not worried about the various factors going with the a and $a^{\dagger}$. This is not very difficult. Let us take the first diagram. We find:

$$\frac{1}{\sqrt{2V p_0}}e^{ipy'}\frac{1}{\sqrt{2V q_0}}e^{iqy'}$$
$$\cdot\left(\frac{1}{\sqrt{2V k_0}}e^{-iky'}\frac{1}{\sqrt{2V k_0}}e^{iky}\right)$$
$$\cdot\frac{1}{\sqrt{2V p_0'}}e^{-ip'y}\frac{1}{\sqrt{2V q_0'}}e^{-iq'y}.$$

This is for $y_0 > y_0'$. For $y_0' > y_0$ the order of the $\mathcal{H}$ is inversed, which is of no consequence to the π part, but now the σ starts in the point y and ends in y'. All together we get:

$$\int d_4 y d_4 y' \frac{(i)^2 g^2}{4V^2 \sqrt{p_0 q_0 p_0' q_0'}} \cdot e^{ipy'} e^{iqy'} e^{-ip'y} e^{-iq'y}$$

$$\cdot \left[\theta\left(y_0 - y_0'\right) \sum_k \frac{1}{2Vk_0} e^{iky - iky'} + \right.$$

$$\left. \theta\left(y_0' - y_0\right) \sum_k \frac{1}{2Vk_0} e^{iky' - iky} \right].$$

Let us now first work out the expression in square brackets. The sum over k may be written as an integral over $d_3 k$, and by methods as described before we may rewrite the whole in terms of a four-dimensional integral:

$$\sum_k \frac{1}{2Vk_0} e^{ik(y-y')} \rightarrow \frac{1}{(2\pi)^3} \int d_4 k\, e^{ik(y-y')} \theta(k_0) \delta\left(k^2 + m^2\right).$$

This function is denoted by $\Delta^+(y-y')$. Similarly the second term, differing only by the interchange of y and y':

$$\sum_k \frac{1}{2Vk_0} e^{ik(y'-y)} \rightarrow \frac{1}{(2\pi)^3} \int d_4 k\, e^{ik(y'-y)} \theta(k_0) \delta\left(k^2 + m^2\right)$$

$$= \frac{1}{(2\pi)^3} \int d_4 k\, e^{ik(y-y')} \theta(-k_0) \delta\left(k^2 + m^2\right) \equiv \Delta^-(y - y').$$

In the last step we replaced k by $-k$. One has evidently:

$$\Delta^+(z) = \Delta^-(-z).$$

The combination

$$\Delta_F(y - y') = \theta\left(y_0 - y_0'\right) \Delta^+(y - y') + \theta\left(y_0' - y_0\right) \Delta^-(y - y')$$

is called the **propagator** of the σ-field. It can be worked out easily using also a Fourier expression for the θ function. One has:

$$\theta(z) = \frac{1}{2\pi i} \int_{-\infty}^{\infty} d\tau \frac{e^{i\tau z}}{\tau - i\epsilon}, \quad \lim \epsilon \rightarrow 0,\ \epsilon > 0.$$

Exercise 3.4 Check this equation by considering the poles of the integrand in the complex τ plane. Add an integral over a large half circle to make a closed contour; take this circle either in the upper or lower τ plane depending on the sign of z such that the exponential becomes very small on the circle.

With this expression we have:

$$\Delta_F(z) = \frac{1}{(2\pi)^4 i} \int d_4 k \, d\tau \, e^{ikz + i\tau z_0}$$

$$\left\{ \frac{\theta(k_0)\delta\left(k^2 + m^2\right)}{\tau - i\epsilon} + \frac{\theta(-k_0)\delta\left(k^2 + m^2\right)}{-\tau - i\epsilon} \right\}.$$

The trick is to get τ out of the exponential. This may be achieved by a change of variable for the k_0 integration. We take:

$$k_0 = k_0' + \tau.$$

Note that $kz = \vec{k}\vec{z} - k_0 z_0$. We so find:

$$\Delta_F(z) = \frac{1}{(2\pi)^4 i} \int d_4 k \int d\tau \, e^{ikz}$$

$$\cdot \left[\frac{\theta(k_0 + \tau)\delta\left(\vec{k}^2 - (k_0 + \tau)^2 + m^2\right)}{\tau - i\epsilon} + \frac{\theta(-k_0 - \tau)\delta\left(\vec{k}^2 - (k_0 + \tau)^2 + m^2\right)}{-\tau - i\epsilon} \right].$$

where we renamed k_0' to k_0. Next we do the τ integral. The argument of the δ-functions is zero if

$$\tau + k_0 = \pm\sqrt{\vec{k}^2 + m^2} \equiv \pm\sqrt{}.$$

The θ functions select the $+$ root for the first term and the $-$ root for the second. The argument of the δ-function can be rewritten

$$\vec{k}^2 - (k_0 + \tau)^2 + m^2 = \left(\sqrt{} - (k_0 + \tau)\right) \cdot \left(\sqrt{} + (k_0 + \tau)\right).$$

Remember again

$$\delta(ab) = \frac{1}{|a|}\delta(b) + \frac{1}{|b|}\delta(a).$$

We find:

$$\Delta_F(z) = \frac{1}{(2\pi)^4 i} \int d_4 k \, e^{ikz}$$

$$\cdot \frac{1}{2\sqrt{\vec{k}^2 + m^2}} \left[\frac{1}{-k_0 + \sqrt{} - i\epsilon} + \frac{1}{k_0 + \sqrt{} - i\epsilon} \right]$$

$$= \frac{1}{(2\pi)^4 i} \int d_4 k \, e^{ikz} \frac{1}{k^2 + m^2 - i\epsilon'}, \quad \lim \epsilon' \to 0,$$

where now $k^2 = \vec{k}^2 - k_0^2$.

Exercise 3.5 Verify this. Give the relation between ϵ' in the last equation with ϵ in the previous line, to see that $\epsilon' > 0$ if $\epsilon > 0$.

The complete expression for the diagram considered is:

$$\frac{-g^2}{(2\pi)^4 i V^2 \sqrt{16 p_0 q_0 p_0' q_0'}}$$

$$\cdot \int d_4 k \int d_4 y d_4 y' \frac{1}{k^2 + m^2 - i\epsilon}$$
$$\exp \left(ipy' + iqy' - ip'y - iq'y + ik(y - y') \right).$$

Both y and y' occur only in the exponents, and the integrals can be done. Using

$$\delta(a) = \frac{1}{2\pi} \int_{-\infty}^{\infty} dx \, e^{iax}$$

we find

$$\frac{-g^2}{(2\pi)^4 i V^2 \sqrt{16 p_0 q_0 p_0' q_0'}} \cdot (2\pi)^8 \int d_4 k \, \delta_4(p + q - k)$$

$$\delta_4(k - p' - q') \frac{1}{k^2 + m^2 - i\epsilon}.$$

The integral over k can be done:

$$\frac{-g^2 (2\pi)^8}{(2\pi)^4 i V^2 \sqrt{16 p_0 q_0 p_0' q_0'}} \frac{\delta_4 (p + q - p' - q')}{(p + q)^2 + m^2 - i\epsilon}.$$

From the above calculation we can see how things go in general. Write down all possible diagrams, and then for any diagram write down the correct factors. As much as possible factors relating to permutations should be absorbed into some easy rules. This is not always possible, but in most cases that we will meet there is really not much of a problem.

First, the combinatorial factor relating to there being two pion lines in a vertex that can be interchanged, is easily taken care of by including a factor of 2 in the vertex. The factor of two relating to the symmetry in y, y' interchange cancels against the factor

1/2! in front of the second order term of the S-matrix expansion. After this we have three essentially different diagrams left:

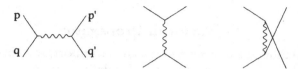

The contribution due to the first diagram has been computed already. The contributions due to the second and third diagram can be computed likewise; one can easily write them down using the Feynman rules for the theory that we are considering here. Mainly, replace $(p+q)^2 = -s$ by $(p-p')^2 = -t$ or $(p-q')^2 = -u$, where s, t and u are the Mandelstam variables. They are not independent: $s+t+u = 4M^2$. Reminder: p' and q' are outgoing. Here are the Feynman rules for the case $\mathcal{H}(x) = g\pi^2(x)\sigma(x)$:

1. To every incoming or outgoing π or σ of momentum p corresponds a factor $1/\sqrt{2Vp_0}$.
2. To every vertex corresponds a factor $2i(2\pi)^4 g\delta_4(\ldots)$. Note: 2. for two pion lines, i from the original equation for the S-matrix, $(2\pi)^4$ from the integral giving the δ-function, and g as found in the interaction Hamiltonian.
3. To every propagator corresponds a factor

$$\frac{1}{(2\pi)^4 i} \int d_4 k \, \frac{1}{k^2 + m^2 - i\epsilon}$$

For the pion field one has M^2 instead of m^2.

Exercise 3.6 Write down the expressions corresponding to the last two diagrams.

Many of the propagator integrals can usually be done, thereby getting rid of δ-functions due to the vertices. The general rule is that one $\delta_4(\ )$ remains, assuring that the sum of incoming momenta equals the total of the outgoing momenta, thus guaranteeing conservation of energy and momentum in any process. In the first non-trivial order (as we are considering here) no momentum integral remains. In the next order one four-dimensional integral remains non-trivial, and in every next order there is one more four-integral. This is what makes it so hard to do higher order calculations.

3.7 Feynman Propagator

This is the moment to reflect on the most important central quantity, the Feynman propagator Δ_F. Its definition was:

$$\Delta_F\left(y - y'\right) = \theta\left(y_0 - y_0'\right)\Delta^+\left(y - y'\right) + \theta\left(y_0' - y_0\right)\Delta^-\left(y - y'\right).$$

The functions Δ^+ and Δ^- are:

$$\Delta^{\pm}(y - y') = \frac{1}{(2\pi)^3}\int d_4k\, e^{ik(y-y')}\theta(\pm k_0)\delta\left(k^2 + m^2\right).$$

In words one may understand the Feynman propagator as follows:

If the time y_0 is larger than the time y_0' then this propagator equals a function containing plane waves for a particle of positive energy on mass shell. We can literally say that if the time $y_0 > y_0'$ then the Feynman propagator represents a physical particle moving from the space-time point y' to the space-time point y. In fact, the exponential is nothing else but the wave function for a plane wave for a particle leaving y' multiplied by the wave function for a particle of the same mass and momentum arriving at y. This product is the overlap of these functions, something that relates to the probability for this to happen. The total propagator is obtained when integrating over all possible physical momenta (positive energy, on mass shell). If the time $y_0' > y_0$ then the particle moves in the opposite direction.

There is a causality idea in there: energy moves from the earlier point to the later. There is another feature: the probability for this to happen must not be negative, which is embodied in the sign of the $\Delta^{\pm}$. Indeed, having a theory with $\Delta^{\pm}$ as above but with a $-$ sign in front would give rise to negative probabilities. This then is the physical content of the Feynman propagator.

The appearance of the θ functions in Δ_F and $\Delta^{\pm}$ thus relates closely to physical concepts. It turns out that these same θ functions are crucial for the study of unitarity of the S-matrix, i.e., conservation of probability, which will be done in another chapter. And the sign of the $\Delta^{\pm}$ relates to the sign of probability: if it happens to be minus for some particle, then that particle better be a ghost, meaning that the sum total of its effects must somehow cancel. Things like that happen in gauge theories, where then the symmetry of the theory guarantees the necessary cancellations.

3.8 Scattering Cross Section

Before we go further with the theory we will connect the results obtained so far with conventional quantum mechanics. In fact, we would like to reproduce the quantum mechanics of the hydrogen atom, and Coulomb scattering. Thus think now of a system of three particles (electron, proton, photon). The electron and proton have the same interaction (apart from a sign) with the e.m. field.

In analogy with this we introduce a new particle in addition to the π and σ, and we will call it P. Apart from the spin, which one usually neglects in first approximation, this P is to play the role of the proton. It interacts with the σ in the same way as the π (the electron). Thus the interaction Hamiltonian becomes:

$$\mathcal{H} = g\pi^2\sigma - gP^2\sigma.$$

The minus sign reflects the fact that the proton charge is opposite to the electron charge.

Exercise 3.7 Verify the following equations of motion:

$$\left(\square - M^2\right)\pi(x) = -2g\pi\sigma, \quad \left(\square - m^2\right)\sigma = -g\pi^2 + gP^2,$$

$$\left(\square - M_p^2\right)P(x) = +2gP\sigma.$$

Use as much as possible previous results, as well as the fact that the P field commutes with all π and σ fields, including time derivatives of these fields.

The Feynman rules are as before, except we now have an extra particle, the P, to be denoted by a broken line. There is also a new vertex, showing the σ–P coupling.

Next we consider πP scattering:

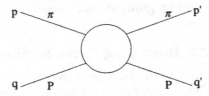

In this case, to second order, we wind up with only one diagram:

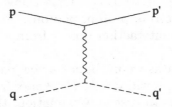

and the corresponding expression for the S-matrix element, or amplitude, is:

$$< S >= \frac{(2g)(-2g)i^2(2\pi)^8}{(2\pi)^4 i V^2 \sqrt{16 p_0 q_0 p_0' q_0'}} \cdot \frac{\delta_4\, (p + q - p' - q')}{(p - p')^2 + m^2 - i\epsilon}$$

where m is the σ-mass.

To obtain a cross section we must take the absolute value squared of this expression, which is not immediately clear because of the δ-function. To get around this we first go back to finite volume V, which amounts to the replacement

$$\delta_3\,(\vec{p} + \vec{q} - \vec{p}' - \vec{q}') \rightarrow \frac{V}{(2\pi)^3} \delta_{\vec{p}+\vec{q},\vec{p}'+\vec{q}'}$$

Thus,

$$(\delta_3(\ ))^2 \rightarrow \frac{V^2}{(2\pi)^6} \delta_{\vec{p}+\vec{q},\vec{p}'+\vec{q}'}$$

since there is nothing difficult about squaring a Kronecker-δ, but we now have V^2 instead of V. Recombining one V factor with the δ:

$$(\delta_3(\ ))^2 = \frac{V}{(2\pi)^3}\delta_3(\ ).$$

Now what about the fourth δ-function (relating to energy conservation)? Here we must introduce a time interval T. Since with plane waves as we consider here there is really no beginning and end to the scattering process we limit our observations to a time interval T, and will compute the transition probability per unit of time. Essentially now things are entirely the same for time and space, and squaring the fourth δ-function gives us a factor $T/2\pi$.

The transition probability is therefore:

$$|<S>|^2 = \left| \frac{-4ig^2(2\pi)^4}{V^2\sqrt{16p_0q_0p_0'q_0'}} \cdot \right.$$

$$\left. \frac{1}{(p-p')^2 + m^2 - i\epsilon} \right|^2 \cdot \frac{VT}{(2\pi)^4}\delta_4(\ ).$$

We want to compare our results to the classical Rutherford scattering cross section formula, thus we must work this over to a cross section. Imagine the (proton) P to be at rest ($q = 0,0,0,iM_p$) and the (electron) π comes in along the z-axis ($p = 0,0,p_z,ip_0$). We thus have a stream of π coming along the z-axis. Since we have one particle in the whole universe the flux (=number of particles per unit surface per unit of time) is v/V where v is the velocity of the π. This velocity is given by $v = |\vec{p}|/p_0$ (check this in the non-relativistic limit). The cross section is the probability per unit of time, for unit flux, summed over all possible final states:

$$\sigma_{\text{tot}} = \frac{V}{(2\pi)^3} \int d_3p' \frac{V}{(2\pi)^3} \int d_3q' \frac{Vp_0}{|\vec{p}|} \cdot |<S>|^2 \cdot \frac{1}{T}.$$

The integrals with the factors in front are simply the continuum limit of summation over all p' and q'. We so arrive at the equation:

$$\sigma_{\text{tot}} = \int d_3p' d_3q' \frac{V^2}{(2\pi)^6} \frac{Vp_0}{T|\vec{p}|} \cdot \frac{VT}{(2\pi)^4}\delta_4(p+q-p'-q')$$

$$\cdot \left| \frac{-4ig^2(2\pi)^4}{V^2\sqrt{16p_0q_0p_0'q_0'}} \cdot \frac{1}{(p-p')^2+m^2-i\epsilon} \right|^2$$

$$= \frac{16g^4}{(2\pi)^2} \cdot \frac{1}{4p_0q_0} \int \frac{d_3p'}{2p_0'} \int \frac{d_3q'}{2q_0'} \cdot \frac{p_0}{|\vec{p}|} \frac{\delta_4(p+q-p'-q')}{|(p-p')^2+m^2-i\epsilon|^2}.$$

The integral over q' can be done, using up three of the δ-functions:

$$\sigma_{\text{tot}} = \frac{16g^4}{(2\pi)^2} \cdot \frac{1}{4p_0q_0} \int \frac{d_3p'}{4p_0'q_0'} \cdot \frac{p_0}{|\vec{p}|} \frac{\delta(p_0+q_0-p_0'-q_0')}{|(p-p')^2+m^2-i\epsilon|^2}$$

where $q_0' = \sqrt{\vec{q}'^2 + M^2}$ with $\vec{q}' = \vec{p} + \vec{q} - \vec{p}'$.

We now make the non-rel-
ativistic approximation and
also the no-recoil approximation, which is the approxima-

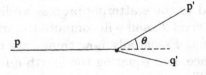

tion that the P mass M_p is
much heavier than the mass M of the π. First introduce polar
coordinates for $\vec{p}'$, the outgoing π momentum:

$$\int d_3 p' = \int d\Omega \int |y|^2 dy$$

where $y = |\vec{p}'|$. Conservation of momentum tells us that q_0', the
energy of the outgoing P, is given by

$$q_0' = \sqrt{\vec{Q}^2 + M_p^2},$$

with $Q = p + q - p'$, and thus

$$\vec{Q} = \vec{p} + \vec{q} - \vec{p}' = \vec{p} - \vec{p}'$$

because $\vec{q} = 0$, as the initial proton is at rest. The quantity $\vec{Q}$ is
called the **momentum transfer**. It is the amount of momentum
given by the π to the P.

If M_p is very large we may approximate:

$$q_0' = M_p + \frac{\vec{Q}^2}{2M_p} + \cdots$$

The no-recoil approximation is to neglect the term $\vec{Q}^2/2M_p$ with
respect to M_p. Thus to this approximation the proton remains at
rest, $q_0' = M_p$. The expression for σ becomes now:

$$\sigma_{\text{tot}} = \frac{g^4}{4\pi^2 p_0 q_0} \int d\Omega \int \frac{y^2 dy}{p_0' q_0'} \frac{p_0}{|\vec{p}|} \frac{\delta(p_0 - p_0')}{|Q^2 + m^2 - i\epsilon|^2}.$$

Now $y = |\vec{p}'|$, thus $p_0' = \sqrt{y^2 + M^2}$. It follows:

$$\frac{dp_0'}{dy} = \frac{1}{2} \cdot \frac{1}{\sqrt{y^2 + M^2}} \cdot 2y = \frac{y}{p_0'},$$

or

$$y \, dy = p_0' \, dp_0' \quad (\text{i.e., } |\vec{p}'| \, |d|\vec{p}'| = p_0' \, dp_0').$$

Furthermore, the δ-function assures us that $p_0 = p_0'$, thus
$\sqrt{\vec{p}^2 + M^2} = \sqrt{\vec{p}'^2 + M^2}$, thus $|\vec{p}| = |\vec{p}'|$. We arrive at

$$\sigma_{\text{tot}} = \frac{g^4}{4\pi^2 p_0 q_0} \int d\Omega \int \frac{dp_0'}{p_0' q_0'} \frac{y p_0' p_0}{|\vec{p}|} \frac{\delta(p_0 - p_0')}{|\cdots|^2}.$$

Using that $y = |\vec{p}'| = |\vec{p}|$ and doing the now trivial p'_0 integration:

$$\sigma_{tot} = \frac{g^4}{4\pi^2 p_0 q_0} \int d\Omega \frac{p_0}{q'_0} \frac{1}{|Q^2 + m^2 - i\epsilon|^2}.$$

It should be noted that in the no-recoil approximation $Q_0 \ll |\vec{Q}|$. This follows because Q_0 is the difference between the initial and final P energy:

$$Q_0 = q'_0 - q_0 = \sqrt{\vec{Q}^2 + M_p^2} - M_p \simeq \frac{\vec{Q}^2}{2M_p}.$$

Therefore $Q^2 = \vec{Q}^2$ in good approximation. Replacing, non-relativistically, q_0 by M_p and p_0 by M we have the final result:

$$\sigma_{tot} = \frac{g^4}{4\pi^2 M_p^2} \int d\Omega \frac{1}{\left(\vec{Q}^2 + m^2\right)^2}.$$

We have omitted the $i\epsilon$ in the propagator, because both $\vec{Q}^2$ and m^2 are positive, so the infinitesimal ϵ is of no relevance here.

If we take the σ mass to be zero we have

$$\sigma_{tot} = \frac{g^4}{4\pi^2 M_p^2} \int d\Omega \frac{1}{\vec{Q}^4}.$$

Now $\vec{Q} = \vec{p} - \vec{p}'$, and

$$\vec{Q}^2 = 2 |\vec{p}|^2 (1 - \cos\theta) = 4 |\vec{p}|^2 \sin^2\frac{\theta}{2}$$

where we used that $|\vec{p}| = |\vec{p}'|$, and θ is the angle that the outgoing π makes with the z-axis (the direction of the incoming π). Thus

$$\sigma_{tot} = \frac{g^4}{4\pi^2 M_p^2} \cdot \int d\Omega \frac{1}{16|\vec{p}|^4 \sin^4(\theta/2)}$$

$$= \frac{g^4}{64\pi^2 M_p^2 M^4} \int d\Omega \frac{1}{v^4 \sin^4(\theta/2)}$$

$$= \frac{2\pi g^4}{64\pi^2 M_p^2 M^4} \int \frac{\sin\theta d\theta}{v^4 \sin^4(\theta/2)},$$

which is indeed the angular distribution as given by the Rutherford scattering formula. In here v is the velocity of the initial π, with $|\vec{p}| = Mv$ as usual, and $d\Omega = \sin\theta d\theta d\phi$.

The above is also the result of the calculation using the

Schrödinger equation for a potential $1/r$. In that case the result amounts essentially to computing the Fourier transform of the potential $1/r$:

$$\int d_3 r\, e^{i\vec{Q}\vec{r}} \frac{1}{r} = 2\pi \int_0^\infty r^2 dr \int_{-1}^1 dz \frac{e^{iQrz}}{r}$$

$$= \frac{2\pi}{\vec{Q}^2} \int_0^\infty x^2 dx \int_{-1}^1 dz \frac{e^{ixz}}{x} = \frac{4\pi}{\vec{Q}^2},$$

with $r = x/Q$ and $Q = |\vec{Q}|$. One uses $\lim_{x\to\infty} e^{\pm ix} = 0$. This $1/Q^2$ corresponds to the propagator

$$\frac{1}{\vec{Q}^2 + m^2}$$

for $m = 0$. If we had not taken $m = 0$ we would have obtained the same result as in the Schrödinger calculation for a potential of the form

$$\frac{1}{r} e^{-mr}$$

Indeed:

$$\int d_3 r \frac{e^{i\vec{Q}\vec{r} - mr}}{r} = \frac{4\pi}{\vec{Q}^2 + m^2}.$$

Such a potential is called a **Yukawa potential**. It is a Coulomb potential $1/r$ combined with a factor that is near 1 for small r, but becomes small if $mr \geq 1$. In nuclei it is observed that forces become weak for distances of the order of 10^{-13} cm. Assume now a Yukawa potential; it follows that the corresponding $m \sim 10^{13}$ cm^{-1}. Multiplying with $\hbar c = 1.9733 \times 10^{-11}$ Mev · cm we find a mass of about 200 MeV. Thus Yukawa guessed in 1940 the existence of a particle of that mass. We now believe this to be the pion, with mass ~ 140 MeV.

3.9 Lifetime

In this section we will consider another application of the theory developed so far, namely the calculation of a decay rate, or a lifetime for an unstable particle.

In principle we have here a contradiction. An unstable particle lives a finite time, and therefore it is impossible to be present at either minus or plus infinite time. In other words, an unstable particle will not occur in in- or out-states. Yet we will do as if it can be in an in-state and then calculate its decay probability. The full justification for that requires a complete treatment of unstable particles, which we will not do here.

Going back to the interaction Hamiltonian described before, $\mathcal{H} = g\pi^2\sigma$, we will now consider the case that the σ-particle is heavier than two pions, so that it normally will decay. The decay probability per unit time is called the **decay rate**, and the inverse of the decay rate is the **lifetime**.

The process of interest has initially a σ and finally two pions. We thus must consider

$$< pq \mid S \mid k >$$

where k denotes the momentum of the initial σ and p and q are the momenta of the final pions.

In lowest order there is one non-vanishing diagram, see figure. The corresponding expression is:

$$< S > = \frac{i(2\pi)^4 2g}{\sqrt{8p_0 q_0 k_0 V^3}} \delta_4(k - p - q).$$

The transition probability is:

$$| < S > |^2 = \frac{(2\pi)^8 4g^2}{8p_0 q_0 k_0 V^3} \frac{V}{(2\pi)^3} \cdot \frac{T}{2\pi} \delta_4(k - p - q).$$

The transition probability per unit of time follows by dividing by T. We must also sum over all final states if we want the total decay rate. This rate is therefore:

$$\Gamma(\sigma \to 2\pi) = \frac{1}{2} \frac{V}{(2\pi)^3} \int d_3 p \frac{V}{(2\pi)^3} \int d_3 q$$

$$\cdot \frac{(2\pi)^8 4g^2}{8p_0 q_0 k_0 V^3} \cdot \frac{V}{(2\pi)^3} \cdot \frac{1}{2\pi} \delta_4(k - p - q)$$

$$= \frac{g^2}{(2\pi)^2 k_0} \int \frac{d_3 p}{2p_0} \int \frac{d_3 q}{2q_0} \delta_4(k - p - q).$$

There is a subtlety here, we have divided by 2. This is because the two pions in the final state are identical, and if we integrate over all momenta we will count double, because the final state with the two pions interchanged is the same state (Bose–Einstein statistics).

To work out the above expression we go to the σ rest system. In this system $k_0 = m$, and $\vec{k} = 0$. This makes the $\vec{q}$ integral trivial, giving $\vec{q} = -\vec{p}$ and therefore $q_0 = p_0$. We get:

$$\Gamma(\sigma \to 2\pi) = \frac{g^2}{4\pi^2 m} \int \frac{d_3 p}{4 p_0^2} \delta(m - 2p_0).$$

We go to polar coordinates. The integral over angles is also trivial giving a factor 4π:

$$\Gamma(\sigma \to 2\pi) = \frac{g^2}{4\pi m} \int \frac{p^2 dp}{p_0^2} \delta(m - 2p_0).$$

Now $p = |\vec{p}|$. Using the relation $p dp = p_0 dp_0$ also this last integral is trivial and we obtain:

$$\Gamma(\sigma \to 2\pi) = \frac{g^2}{8\pi m} \frac{p}{p_0}$$

with $p_0 = \frac{1}{2} m$ and $p = \sqrt{p_0^2 - M^2}$. Thus

$$\Gamma(\sigma \to 2\pi) = \frac{g^2}{8\pi m^2} \sqrt{m^2 - 4M^2}.$$

The lifetime is the inverse of this, $\tau = 1/\Gamma$.

3.10 Numerical Evaluation

Generally one wants a cross section in terms of cm^2, and a lifetime in seconds. We have used $\hbar = c = 1$, and will express everything else in MeV. The cross section will have the dimension of $(MeV)^{-2}$, the decay rate is of dimension MeV (and lifetime $(MeV)^{-1}$). To go to cm^2 the cross section must be multiplied by $(\hbar c)^2 = (1.97327 \times 10^{-11} \, MeV \cdot cm)^2$. To go from MeV to sec^{-1} the decay rate must be divided by $\hbar = 6.582122 \times 10^{-22} \, MeV \cdot sec$, and the lifetime is thus $\hbar/\Gamma$. Note that in the examples above the coupling constant g has the dimension MeV.

3.11 Schrödinger Equation, Bound States

In this section we will show how the Schrödinger equation of ordinary quantum mechanics is contained in the theory. It is not difficult to obtain the Dirac equation also, but here we want to avoid the complications of spinors and γ-matrices. They are not essential to the argument.

It is often thought that bound state problems are not part of the S-matrix theory, that is that the theory developed so far is not complete. In actual fact the connection is quite simple, and understanding it helps in attacking problems where the approximations of ordinary quantum mechanics are not sufficient. It will also become clear how bound states in general fit into the S-matrix scheme.

The interaction to be used is the well-known σ–π and σ–P interactions used first to derive the Rutherford scattering cross section. We include factors M and M_P in the vertices from the start, which makes the coupling constant dimensionless. The interaction Hamiltonian is

$$\mathcal{H}_i = gM\sigma\pi^2 - gM_P\sigma P^2.$$

This leads to the vertex factors $2Mg$ and $-2gM_P$ for the $\sigma\pi\pi$ and σPP vertices respectively.

The main approximation is in the type of diagrams taken into account. Furthermore, to obtain something like a potential some further approximations must be made, amounting mostly to the non-relativistic approximation.

Consider now π–P scattering. We will make the ladder diagram approximation, which means that we restrict ourselves to diagrams involving σ exchange. That means we consider the series of diagrams shown in the figure. They are called ladder diagrams.

Some examples of ignored diagrams are shown in the next figure.

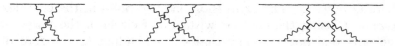

All ladder diagrams are build of the same basic part, namely an exchanged σ particle, a π-propagator and a P-propagator. This part is shown in the figure. There is one 4-dimensional integration associated with this, and the trick is to do the integral over the fourth component. The expression corresponding to the quantities mentioned is:

$$-\frac{g^2\,4MM_P}{(2\pi)^4 i}\int d_4q\,V(q)$$

$$\frac{1}{((p-q)^2 + M^2 - i\epsilon)((k+q)^2 + M_P^2 - i\epsilon)}.$$

The quantity $V(q)$ is the σ-propagator. Remember our convention $d_4q = d_3q\,dq_0$, not involving the imaginary i. We will perform the q_0 integral, but in order to do that we must make some approximations. Obviously, as we know already from previous discussions, the σ-propagator plays the role of a potential, but in the approximation that $q^2 = \vec{q}^{\,2} - q_0^2 \approx \vec{q}^{\,2}$. In the non-relativistic approximation q_0 is of the same order as $\vec{q}^{\,2}/M$, with large M. Thus we ignore q_0 dependence in the σ propagator, not only in $V(q)$, but also in the σ-propagator of the next ladder element. Then all q_0 dependence is contained in the expression shown above, and the integral can be done easily. Of course, also in this expression we will ignore q_0^2 compared to $\vec{q}^{\,2}$.

Assuming that we are in the π-P rest frame we have $\vec{k} = -\vec{p}$. Furthermore we introduce $E_\pi = p_0 - M$ and $E_P = k_0 - M_P$, and in good approximation, ignoring terms of order E_π^2 and E_P^2 we may write:

$$p^2 + M^2 = \vec{p}^{\,2} - 2ME_\pi\,,\quad k^2 + M_P^2 = \vec{k}^{\,2} - 2M_P E_P = \vec{p}^{\,2} - 2M_P E_P\,.$$

The expression for one element of the ladder becomes:

$$-\frac{g^2\,4MM_P}{(2\pi)^4 i}\int d_3q\,V(q)$$

$$\int \frac{dq_0}{(2p_0 q_0 + \vec{Q}^2 - 2ME_\pi - i\epsilon)(-2k_0 q_0 + \vec{Q}^2 - 2M_P E_P - i\epsilon)}.$$

In here $\vec{Q} = \vec{p} - \vec{q}$. The expression has two poles in the complex q_0-plane. Closing the contour with a half circle in the lower q_0 plane only one pole, namely that corresponding to the zero of the second factor in the denominator contributes. As usual, since the integrand behaves as $1/q_0^2$ for large q_0, we may ignore the contribution of the half circle, thus the result is unchanged by the addition of this half circle. For this clockwise contour we obtain $-2\pi i$ times the residue of the pole. The residue of the second factor is simple $-1/2k_0$, and putting in the appropriate value of

q_0 in the first factor we obtain:

$$\frac{\dfrac{-g^2 4MM_P}{(2\pi)^4 i} \displaystyle\int d_3q\, V(q)}{(p_0 + k_0)\vec{Q}^2 - 2k_0 E_\pi M - 2p_0 E_P M_P - i\epsilon},$$

At this point we may actually set $p_0 = M$ and $k_0 = M_P$. The result is:

$$\frac{-2i\pi g^2}{(2\pi)^4 i} \int d_3q\, V(q)\, \frac{1}{\frac{1}{2\mu}(\vec{p} - \vec{q})^2 - E - i\epsilon}.$$

The quantity $\mu = MM_P/(M + M_P)$ is the reduced mass of the two particle π–P system, and E is the total energy of the two particles. To bring this into a form that makes it easy to attach more ladder elements we substitute $\vec{q} = \vec{p} - \vec{q}\,'$, and then we drop the prime:

$$\frac{g^2}{(2\pi)^3} \int d_3q\, \frac{1}{E - \frac{1}{2\mu}\vec{q}^2 + i\epsilon}\, V(\vec{p} - \vec{q}).$$

Now $\vec{q}$ is the three-momentum of the continuing pion line, and ladder elements may be chained. For this kind of situations an equation may be derived by noting a recursion here in case of an infinite ladder. Remember, the full expression contains the diagram without any element (ladder with zero elements), plus a ladder with one element, plus a ladder of two elements etc. This is much like a geometric series, $f(x) = 1 + x + x^2 + x^3 \dots$. One has, obviously, the equation $f(x) = 1 + f(x)x$ from which one solves $f(x) = 1/(1-x)$. We will do precisely that. Let $F(\vec{q}, \vec{p})$ denote the full collection for a situation with incoming three-momentum $\vec{p}$ and outgoing three-momentum $\vec{q}$. There is an integral over $\vec{q}$, but we take that out. The zero ladder is simply 1, which we write as $\int d_3q\, \delta_3(\vec{p} - \vec{q})$, and taking out the integral over $\vec{q}$ that becomes $\delta_3(\vec{p} - \vec{q})$. We then have the following equation:

$$F(\vec{q}, \vec{p}) = \delta_3(\vec{p} - \vec{q}) + \frac{g^2}{(2\pi)^3} \int d_3k\, \frac{1}{E - \frac{1}{2\mu}\vec{q}^2 + i\epsilon}\, V(\vec{k} - \vec{q})\, F(\vec{k}, \vec{p}).$$

This corresponds to the figure, where the shadowed box stands for the full collection of ladder diagrams including the zero ladder. As usual, the sequence in the formula corresponds to reading the diagrams from end to beginning. That this equation is correct can be verified by iteration. Writing for $F(\vec{k},\vec{p})$ the lowest order approximation, i.e., $\delta_3(\vec{p}-\vec{k})$, one obtains the correct one element ladder, inserting that again gives the two element ladder etc.

To proceed further we write the Fourier transform of this equation. We claim that the following equation is the desired Fourier transform of the above:

$$f(\vec{r},\vec{p}) = e^{i\vec{p}\vec{r}} + \int d_3r'\, G(\vec{r}',\vec{r})\, V(\vec{r}')\, f(\vec{r}',\vec{p})\,,$$

where

$$f(\vec{r},\vec{p}) = \int d_3q\, F(\vec{q},\vec{p})\, e^{i\vec{q}\vec{r}}$$

$$V(\vec{r}) = \frac{g^2}{(2\pi)^3} \int d_3Q\, V(\vec{Q})\, e^{-i\vec{Q}\vec{r}}$$

$$G(\vec{r}',\vec{r}) = \frac{1}{(2\pi)^3} \int d_3q\, \frac{1}{E - \frac{1}{2\mu}\vec{q}^2}\, e^{i\vec{q}(\vec{r}-\vec{r}')}\,.$$

To see that this is the same first integrate over $\vec{r}'$, which gives $(2\pi)^3\delta(\vec{Q}-\vec{k}+\vec{q})$. Then the integral over $\vec{Q}$ may be done and one obtains the equation as before but multiplied with $\exp(i\vec{r}\vec{q})$ and integrated over $\vec{q}$.

If the energy $E = \vec{p}^2/2\mu$, as corresponding to incoming physical (on mass-shell) particles then this equation is the Lippmann–Schwinger equation describing scattering off a potential. By simple insertion it may be seen that the function $f(\vec{r},\vec{p})$ is a solution of the Schrödinger equation:

$$\left(\frac{1}{2\mu}\Delta + E\right) f(\vec{r},\vec{p}) = V(\vec{r})f(\vec{r},\vec{p})\,,$$

where $\Delta = \partial_1^2 + \partial_2^2 + \partial_3^2$.

Now about bound states. We will not actually demonstrate all the calculational details, but roughly sketch what happens. The Schrödinger equation may have discrete solutions with negative E, denoted by E_n. To these will correspond solutions that behave around these energies as $1/(E-E_n-i\epsilon)$, that is apart from a factor just like a propagator for a particle at rest (remember, we were in the π–P rest system). Thus the bound states appear as propagating particles inside the S-matrix, and to maintain unitarity one must then also add such states as possible in- and out-states. An example of such an occurrence is shown in the figure. The circles stand for an unspecified collection of diagrams.

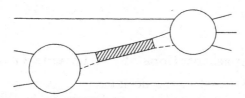

The above argument will be understood in detail when considering the proof of unitarity of the S-matrix, something investigated in a later chapter. Briefly, if a particle can occur somewhere in the middle, then it also can appear in the beginning or at the end, unless it is unstable. Whether a particle is unstable may be deduced from the width of the resonance curve as one passes through the energy region. Here that aspect will not be discussed.

4
Particles with Spin

4.1 Representations of the Lorentz Group

As has been discussed in the beginning, to every Lorentz transformation there corresponds a transformation in Hilbert space. For example, let there be a Lorentz transformation L such that a vector p_μ is transformed into p'_μ:

$$p'_\mu = L_{\mu\nu} p_\nu.$$

Corresponding to this Lorentz transformation there will be a transformation U in Hilbert space such that U applied to a vector corresponding to a state with one (or more) particles with momentum p is transformed into the vector corresponding to the state with one (or more) particles with momentum p' :

$$|\, p' > \propto \ U \,|\, p > .$$

To show that this U is the one corresponding to L we will write $U(L)$. Note that we used the $\propto$ sign rather then an $=$ sign. This is to indicate our somewhat unrealistic discussion. Having chosen a volume and obtaining a discrete set of momentum values, we have violated Lorentz invariance, and it is in general not even true that the Lorentz transform of a momentum is again one of the discrete values. Also the normalization to one particle in the volume V is not invariant. But we simply do not want to get involved in those details.

In principle, for scalar particles the matrix U is known. For instance:

$$U(L) \,|\, p_1 p_2 p_3 \ldots > \propto \, |\, p'_1 p'_2 p'_3 \ldots >$$

with $p'_1 = L p_1$, $p'_2 = L p_2$, etc.

The matrix U just permutes the basis vectors in Hilbert space (assuming the basis vectors correspond to sharp momentum states). With proper normalization, in the limit of infinite volume, U is unitary, that is $U^\dagger = U^{-1}$.

However, while U is unitary it is not really like a rotation. For a rotation one could expect to get back near to the unit matrix by applying the matrix sufficiently many times, for instance doing 360 times a rotation over 1° gives the identity. But let there be given some momentum p, with $Lp = p'$ and $p'_0 \neq p_0$. No matter how often applied one never gets back to p. If p corresponds to a particle at rest, then Lp corresponds to that particle moving with some speed, and LLp to that particle moving faster etc. etc., but $LL \ldots Lp$ will never be like a particle approximately at rest. Again, this property reflects the fact that the Lorentz group is not compact. It is a known fact in group theory that unitary representations of non-compact groups are of infinite dimension, which is what we have with U here.

It is unfortunate that the choice of periodic boundary conditions is not Lorentz invariant. However, since volume is preserved under a translation there is no difficulty with translation invariance. It is instructive to consider that for a moment.

As argued before, a translation over a four-vector a generates a transformation in Hilbert space:

$$| p > \rightarrow e^{iap} | p >, \quad | p, q, \ldots > \rightarrow e^{ia(p+q+\ldots)} | p, q, \ldots > .$$

To us the important quantities are the fields. These fields are operators, that is matrices in Hilbert space. A transformation in Hilbert space will induce a transformation of those fields. To see what they do, consider a matrix-element of the scalar field $\sigma(x)$ in between a one particle state and the vacuum, $< 0 | \sigma(x) | p >$. One of the terms in the σ field (the one containing the annihilation operator for momentum $\vec{p}$) will transform the state $| p >$ into the vacuum state, and the result is the coefficient found in that term:

$$< 0 | \sigma(x) | p > = \frac{1}{\sqrt{2p_0 V}} e^{ipx} .$$

Now perform a translation. Then the state $| p >$ obtains a factor $\exp(ipa)$ while the vacuum state remains invariant. This gives

$$< 0 | \sigma(x) | p > \rightarrow \frac{1}{\sqrt{2p_0 V}} e^{ipx+ipa} = < 0 | \sigma(x + a) | p > .$$

We thus obtain the expected behaviour of the σ field under translations. This then is the point. Some transformation in Hilbert space, not at all appearing like a translation, induces a transformation of the fields that is truly like a conventional translation. Also note that the transformation in Hilbert space is obviously unitary, all states are multiplied by some phase factor. This is the way things are supposed to work in general. Thus Lorentz transformations in Hilbert space induce transformations of the fields. It is really these latter transformations that are of interest to us, and they will be crucial with respect to the Lorentz invariance of the final theory in terms of Feynman rules. Note that we could work backward: if the behaviour of the fields is known then the transformation in Hilbert space can be deduced.

For scalar particles one has to deal only with states of well defined momentum, but no further degrees of freedom. For particles with spin this is different. An electron for example has spin 1/2, and there are in any given situation two electrons: electrons with spin up or electrons with spin down. Furthermore, there are positrons of spin 1/2 that also can occur with spin up or spin down. Hilbert space must be enlarged. We now have:

$| \, 0 >$ No electron or positron.

$| \, \vec{p} \uparrow >$ One electron of momentum $\vec{p}$ with spin up.

$| \, \vec{p} \downarrow >$ One electron of momentum $\vec{p}$ with spin down.

$| \, \underline{\vec{p}} \uparrow >$ One positron of momentum $\vec{p}$ with spin up.

$\vdots$

$| \, \vec{p}_1 \uparrow, \vec{p}_2 \downarrow, \underline{\vec{p}}_3 \uparrow \ldots >$

The problem now is what happens to these states if a Lorentz transformation is performed. This is easy for some cases, for instance, making a rotation over 180° around the x-axis transforms spin up into spin down and vice versa. Actually, this also may change the direction of the momentum, so this transformation transforms $|\vec{p} \uparrow >$ into $|\vec{p}\,' \downarrow >$ where $\vec{p}\,'$ is obtained from $\vec{p}$ by the rotation around the x-axis. Thus things get complicated here. The really simple case is for particles at rest $(p = 0, 0, 0, im)$ because then, under rotations in three-dimensional space the momentum p is unchanged. In general what we will have is that under a Lorentz transformation two things happen: first one transforms from one momentum to some other, and secondly the spin states are rotated into each other. There is also the problem of what is spin up and what is spin down, because what is up and what is down is not necessarily the same for systems moving with some speed with respect to each other.

We will not try to deduce in the most general way what is and what is not possible. Instead we will simply describe the situation for the cases that we need, and that is first of all spin 1/2 particles with mass.

With respect to the fields one imagines this as follows. A particle with four states is really nothing else but four different particles, and the fact that a Lorentz transformation may turn one into another is another matter. Correspondingly, there will be four fields, to correspond to these four particles. Lorentz transformations transform these fields into each other. We are concentrating on the transformation of these fields.

We will therefore construct a representation of the Lorentz group in an abstract four-dimensional space, not to be confused with ordinary space-time. The transformations in that space will be the transformations of the fields, not really the transformations in Hilbert space even if there is a tight relationship. In other words, we are skipping a step. Basically we take the Lorentz transformations in Hilbert space for granted, and describe the transformations that they induce in the fields. And as a matter of fact, once we are through, we could deduce the form of the transformations in Hilbert space from the transformations of the fields. But we are not really interested in that.

The general Lorentz transformation is of the form:

$$L = e^{\frac{1}{2}\alpha^{\mu\nu} K_{\mu\nu}}$$

where the $\alpha^{\mu\nu}$ are certain numbers specifying the Lorentz transformation, and the $K_{\mu\nu}$ are matrices given before.

As a first step we now specify a set of matrices X in a four-dimensional space (not ordinary space-time) such that to every L there corresponds an X, and to the product of two L corresponds the product of the X. Thus:

$$L_1 \longrightarrow X(L_1)$$
$$L_2 \longrightarrow X(L_2)$$
$$L_3 \longrightarrow X(L_3)$$

and if $L_3 = L_1 L_2$ then $X(L_3) = X(L_1)X(L_2)$. While the matrices X are also four by four matrices, they are very different from the L. To describe them we will have to go through a number of steps. First we define four matrices, called γ-matrices:

$$\gamma^1 = \begin{bmatrix} 0 & 0 & 0 & -i \\ 0 & 0 & -i & 0 \\ 0 & i & 0 & 0 \\ i & 0 & 0 & 0 \end{bmatrix} \qquad \gamma^2 = \begin{bmatrix} 0 & 0 & 0 & -1 \\ 0 & 0 & 1 & 0 \\ 0 & 1 & 0 & 0 \\ -1 & 0 & 0 & 0 \end{bmatrix}$$

$$\gamma^3 = \begin{bmatrix} 0 & 0 & -i & 0 \\ 0 & 0 & 0 & i \\ i & 0 & 0 & 0 \\ 0 & -i & 0 & 0 \end{bmatrix} \qquad \gamma^4 = \begin{bmatrix} 1 & 0 & 0 & 0 \\ 0 & 1 & 0 & 0 \\ 0 & 0 & -1 & 0 \\ 0 & 0 & 0 & -1 \end{bmatrix}$$

$$\gamma^5 = \begin{bmatrix} 0 & 0 & -1 & 0 \\ 0 & 0 & 0 & -1 \\ -1 & 0 & 0 & 0 \\ 0 & -1 & 0 & 0 \end{bmatrix}$$

where $\gamma^5 = \gamma^1 \gamma^2 \gamma^3 \gamma^4$. These γ-matrices have the following properties:

$$(\gamma^\mu)^\dagger = \gamma^\mu \qquad (\dagger = \text{complex conj.} + \text{reflection})$$
$$(\gamma^\mu)^2 = I$$
$$\gamma^\mu \gamma^\nu + \gamma^\nu \gamma^\mu = 2\delta_{\mu\nu} I$$

where I is the unit matrix, and μ, ν can be any of the values 1...5.

Exercise 4.1 Verify these properties.

Secondly we define a set of matrices $\sigma_{\mu\nu}$, with $\mu, \nu = 1, \ldots, 4$:

$$\sigma_{\mu\nu} = \tfrac{1}{4} \left(\gamma^\mu \gamma^\nu - \gamma^\nu \gamma^\mu \right).$$

Note that $\sigma_{\mu\nu} = -\sigma_{\nu\mu}$, and $\sigma_{\mu\nu} = $ if $\mu = \nu$. These matrices σ take the place of the K-matrices when going from the Lorentz transformations to the X-matrices:

$$L = e^{\frac{1}{2}\alpha^{\mu\nu}K_{\mu\nu}} \rightarrow X = e^{\frac{1}{2}\alpha^{\mu\nu}\sigma_{\mu\nu}}.$$

We now have the explicit form of the X-matrices. The question is now if this mapping is maintained for a product of transformations. This is the essential point. For example, let there be a number of Lorentz transformations such that the product is the unit matrix. For sure do we want that the corresponding product of the transformations X is also the unit matrix (possibly apart from a phase factor). It would be quite unacceptable if doing nothing (the Lorentz identity) would result in a change of state in Hilbert space.

Thus, let there be given two Lorentz transformations specified by two sets of constants α and β. The product will be another Lorentz transformation involving constants κ:

$$e^{\frac{1}{2}\alpha^{\mu\nu}K_{\mu\nu}}e^{\frac{1}{2}\beta^{\mu\nu}K_{\mu\nu}} = e^{\frac{1}{2}\kappa^{\mu\nu}K_{\mu\nu}}$$

The κ will be functions of the α and β. In fact, we know from the Campbell–Baker–Hausdorff equation that the commutation relations of the K determine the result, the κ. The commutation relations of the K are given through the structure constants that we determined earlier.

Consider now the corresponding product of X transformations:

$$e^{\frac{1}{2}\alpha^{\mu\nu}\sigma_{\mu\nu}}e^{\frac{1}{2}\beta^{\mu\nu}\sigma_{\mu\nu}} = e^{\frac{1}{2}\lambda^{\mu\nu}\sigma_{\mu\nu}}$$

Like above, the constants λ are determined if we know the commutation relations of the σ. If the commutation relations of the σ have the same form as those for the K, with the same structure constants, then the λ will be equal to the κ, and the above mapping $L \to X$ preserves the structure of the Lorentz group.

Thus we must check that

$$[\sigma_{ab}, \sigma_{cd}] = c^{ij}_{abcd}\,\sigma_{ij}$$

with the same structure constants c that appear in the commutation relations of the K. This is easy, and we leave it to the reader.

Exercise 4.2 Verify this for the commutator of σ_{12} and σ_{13}.

Let us discuss some properties of the X. For Lorentz transformations we know that $L^{-1} = \widetilde{L}$. This is seen also by realizing that the K-matrices are antisymmetric:

$$\widetilde{K}_{ab} = -K_{ab}$$

from which it follows:

$$\widetilde{L} = e^{\frac{1}{2}\alpha^{\mu\nu}\widetilde{K}_{\mu\nu}} = e^{-\frac{1}{2}\alpha^{\mu\nu}K_{\mu\nu}} = L^{-1}$$

as expected. How is this for the X-matrices? The σ-matrices are not antisymmetrical, but they are antihermitian:

$$\left(\sigma_{ab}\right)^{\dagger} = \tfrac{1}{4}\left(\gamma^{a}\gamma^{b} - \gamma^{b}\gamma^{a}\right)^{\dagger}$$

$$= \tfrac{1}{4}\left(\gamma^{b\dagger}\gamma^{a\dagger} - \gamma^{a\dagger}\gamma^{b\dagger}\right)$$

$$= \tfrac{1}{4}\left(\gamma^{b}\gamma^{a} - \gamma^{a}\gamma^{b}\right) = -\sigma_{ab}$$

where we used $(AB)^{\dagger} = B^{\dagger}A^{\dagger}$. If the constants α would have been real then we would have had $X^{\dagger} = X^{-1}$. However, the α are not real, α_{14}, α_{24}, and α_{34} are imaginary. What now?

To deal with this we consider γ^{4}. The following holds:

$$\gamma^{\mu}\gamma^{4} = -\gamma^{4}\gamma^{\mu} \quad \text{if } \mu = 1, 2, 3$$

$$\gamma^{\mu}\gamma^{4} = \gamma^{4}\gamma^{\mu} \quad \text{if } \mu = 4.$$

$$\sigma_{\mu\nu}\gamma^{4} = \gamma^{4}\sigma_{\mu\nu} \quad \text{if } \mu, \nu = 1, 2, 3$$

$$\sigma_{\mu\nu}\gamma^{4} = -\gamma^{4}\sigma_{\mu\nu} \quad \text{if } \mu \text{ or } \nu = 4,$$

and finally:

$$\left(\alpha^{\mu\nu}\sigma_{\mu\nu}\right)^{\dagger}\gamma^{4} = -\gamma^{4}\left(\alpha^{\mu\nu}\sigma_{\mu\nu}\right).$$

The $-$ sign arising from $(\alpha)^{*} = -\alpha$ if μ or $\nu = 4$ is compensated by the $-$ sign from the γ^{4} commutation. We therefore find the property:

$$X^{\dagger}\gamma^{4} = \gamma^{4}X^{-1}.$$

Another more complicated property holds for the X and γ-matrices. Let L be the matrix given by

$$L = e^{\frac{1}{2}\alpha^{\mu\nu}K_{\mu\nu}}$$

to which corresponds the matrix

$$X = e^{\frac{1}{2}\alpha^{\mu\nu}\sigma_{\mu\nu}}.$$

Then:

$$X^{-1}\gamma^{\alpha}X = L_{\alpha\beta}\gamma^{\beta} \quad \text{(summation over } \beta\text{)}.$$

This is to be seen as the analogue of the way a momentum p transforms under a Lorentz transformation:

$$p'_{\alpha} = L_{\alpha\beta}p_{\beta}.$$

The proof requires expansion of the exponentials. We will verify the lowest order. It is not difficult to verify all orders, but we leave that to the reader. In lowest order we must check:

$$\left(1 - \tfrac{1}{2}\alpha^{\mu\nu}\sigma_{\mu\nu}\right)\gamma^\alpha \left(1 + \tfrac{1}{2}\alpha^{\lambda\delta}\sigma_{\lambda\delta}\right) = \left(1 + \tfrac{1}{2}\alpha^{ab}K_{ab}\right)_{\alpha\beta}\gamma^\beta.$$

As the reader may observe the main problem is to keep all indices straight.

In lowest order we may ignore the term quadratic in the α in the left hand side, and we get:

$$\gamma^\alpha - \tfrac{1}{2}\alpha^{\mu\nu}\sigma_{\mu\nu}\gamma^\alpha + \tfrac{1}{2}\alpha^{\lambda\delta}\gamma^\alpha\sigma_{\lambda\delta}.$$

Note that the α are numbers while σ and γ are matrices. In this expression we now rename λ and δ to μ and ν, and we obtain:

$$\gamma^\alpha - \tfrac{1}{2}\alpha^{\mu\nu}\left[\sigma_{\mu\nu},\gamma^\alpha\right].$$

The commutation rules between the σ and γ are easy to derive:

$$\gamma^\mu\gamma^\nu\gamma^\alpha = -\gamma^\mu\gamma^\alpha\gamma^\nu + 2\delta_{\nu\alpha}\gamma^\mu$$
$$= \gamma^\alpha\gamma^\mu\gamma^\nu + 2\delta_{\nu\alpha}\gamma^\mu - 2\delta_{\mu\alpha}\gamma^\nu,$$

thus,

$$[\gamma^\mu\gamma^\nu,\gamma^\alpha] = 2\delta_{\nu\alpha}\gamma^\mu - 2\delta_{\mu\alpha}\gamma^\nu,$$

and from this:

$$[\sigma^{\mu\nu},\gamma^\alpha] = \delta_{\nu\alpha}\gamma^\mu - \delta_{\mu\alpha}\gamma^\nu.$$

Therefore:

$$\gamma^\alpha - \tfrac{1}{2}\alpha^{\mu\nu}\left[\sigma_{\mu\nu},\gamma^\alpha\right] = \gamma^\alpha - \tfrac{1}{2}\alpha^{\mu\nu}\left(\delta_{\nu\alpha}\gamma^\mu - \delta_{\mu\alpha}\gamma^\nu\right)$$
$$= \gamma^\alpha - \tfrac{1}{2}\alpha^{\mu\nu}\left(\delta_{\nu\alpha}\delta_{\mu\beta} - \delta_{\mu\alpha}\delta_{\nu\beta}\right)\gamma^\beta.$$

In the last step we introduced a new index β. All that is left now is to see that

$$(K_{ab})_{\alpha\beta} = -\delta_{b\alpha}\delta_{a\beta} + \delta_{a\alpha}\delta_{b\beta},$$

for instance:

$$(K_{12})_{\alpha\beta} = \begin{cases} 1 & \text{if } \alpha = 1, \ \beta = 2 \\ -1 & \text{if } \alpha = 2, \ \beta = 1 \\ 0 & \text{otherwise} \end{cases}$$

which is indeed precisely correct.

The reader may note that the essential property in the above derivation is:

$$[\sigma_{\mu\nu},\gamma^\alpha] = -\left(K_{\mu\nu}\right)_{\alpha\beta}\gamma^\beta.$$

Using this equation it is not very difficult to extend the proof to all orders.

4.2 The Dirac Equation

At this point we have a set of matrices X operating in a four-dimensional space. A particle or anti-particle with spin up or down corresponds to some vector in this space. We first make the connection for particles at rest, and will get the vectors for moving particles by applying a Lorentz transformation. Incidentally, to exhibit the difference between the space in which the X operate from normal space we will call vectors in this X-space spinors.

We define the following assignment for particles at rest:

$$\text{electron, spin up } u^1 = \begin{pmatrix} 1 \\ 0 \\ 0 \\ 0 \end{pmatrix} \qquad \text{electron, spin down } u^2 = \begin{pmatrix} 0 \\ 1 \\ 0 \\ 0 \end{pmatrix}$$

These two spinors are also a solution of the matrix equation:

$$Du = 0$$

with

$$D = \begin{bmatrix} 0 & 0 & 0 & 0 \\ 0 & 0 & 0 & 0 \\ 0 & 0 & 1 & 0 \\ 0 & 0 & 0 & 1 \end{bmatrix}$$

Note that $D = \frac{1}{2}(1 - \gamma^4)$. The factor $1/2$ is of no importance. The equation is now:

$$\left(-\gamma^4 + 1\right) u = 0.$$

For a particle of mass m at rest the four-momentum $p = (0, 0, 0, im)$. Only $p_4 \neq 0$. Therefore we can write also:

$$\left(i\gamma^\mu p_\mu + m\right) u = 0.$$

This is the **Dirac equation**. So far it holds by construction. However, we will now show the following. Suppose we apply a Lorentz transformation. Then the four-momentum p becomes p' with $p' = Lp$. Also the spinor u changes to u' with $u' = X(L)u$. Then u' is again a solution of the Dirac equation with p' instead of p. Proof:

$$\left(i\gamma^\mu p'_\mu + m\right) u' = (i\gamma^\mu L_{\mu\nu} p_\nu + m) Xu.$$

Now multiply this expression from the left with X^{-1}. We get

$$\left(iX^{-1}\gamma^\mu X L_{\mu\nu} p_\nu + m\right) u.$$

It was shown before that

$$X^{-1}\gamma^\mu X = L_{\mu\alpha}\gamma^\alpha.$$

We thus obtain:

$$(iL_{\mu\alpha}\gamma^\alpha L_{\mu\nu}p_\nu + m)\, u.$$

Now $L_{\mu\alpha}L_{\mu\nu} = \tilde{L}_{\alpha\mu}L_{\mu\nu} = \delta_{\alpha\nu}$ because $\tilde{L}L = 1$. The result is:

$$(i\gamma^\nu p_\nu + m)\, u$$

which is zero.

We now need the form of the spinor u for particles not at rest, i.e., after a Lorentz transformation. It follows from the above that it will still be a solution of the Dirac equation, thus we can also find this Lorentz transformed u by solving the Dirac equation. One finds the first two solutions of the table:

$$\times\sqrt{\frac{m+E}{2E}}$$

$u^1(p)\uparrow$	$u^2(p)\downarrow$	$u^3(p)\downarrow$	$u^4(p)\uparrow$
1	0	$\dfrac{-p_3}{m+E}$	$\dfrac{p_1-ip_2}{m+E}$
0	1	$-\dfrac{p_1+ip_2}{m+E}$	$\dfrac{-p_3}{m+E}$
$\dfrac{p_3}{m+E}$	$\dfrac{p_1-ip_2}{m+E}$	-1	0
$\dfrac{p_1+ip_2}{m+E}$	$\dfrac{-p_3}{m+E}$	0	1

$$E = \sqrt{\vec{p}^2 + m^2}$$

The last two columns are the solutions obtained analogously to the first two but now for the equation

$$(-i\gamma^\mu p_\mu + m)u = 0$$

These solutions correspond to anti-particles, the positrons. The details will become clear later. Before we go on we note a number of properties. First define:

$$\bar{u} = u^*\gamma^4$$

Note that the dot product between $\bar{u}$ and a u is Lorentz invariant. For consider two spinors u and v. Let there be some Lorentz transformation L and the corresponding $X(L)$, and spinors u' and v' with $u' = Xu$, $v' = Xv$. We have:

$$\bar{u}'v' = u'^*\gamma^4 v' = (Xu)^*\gamma^4 Xv = u^* X^\dagger \gamma^4 Xv$$

Now remember $X^\dagger \gamma^4 = \gamma^4 X^{-1}$. We get:

$$\bar{u}'v' = u^*\gamma^4 X^{-1} Xv = u^*\gamma^4 v = \bar{u}v.$$

Thus, the "dot" product $\bar{u}v$ is Lorentz invariant. The u given in the table are actually not precisely the spinors obtained by Lorentz transformations from the $(1,0,0,0)$ etc., there is a normalization factor. Indeed, one finds:

$$\bar{u}u = m/E \text{ for } u^1 \text{ and } u^2$$
$$= -m/E \text{ for } u^3 \text{ and } u^4.$$

It follows that for instance u^1 is $\sqrt{m/E}$ times the spinor obtained from $(1,0,0,0)$ by a Lorentz transformation. This normalization has some advantages later on.

In addition to the above:

$$u^*u = 1 \quad \text{for all } u,$$

and the spin-sum equations:

$$\sum_{j=1}^{2} u_\beta^j (\vec{p})\, \bar{u}_\alpha^j (\vec{p}) = \frac{1}{2p_0} (-i\gamma p + m)_{\beta\alpha}$$

$$\sum_{j=3}^{4} u_\beta^j (\vec{p})\, \bar{u}_\alpha^j (\vec{p}) = \frac{1}{2p_0} (-i\gamma p - m)_{\beta\alpha}.$$

Exercise 4.3 Verify these equations.

For completeness here are a few more equations. There are spinor transformations that correspond to space reflection, charge conjugation and time reversal. Apart from possibly a sign they transform spinors to what you would expect. Space reflections change the sign of the three-momentum, but not the spin direction. For space reflection (parity) the transformation matrix is $P = \gamma^4$, with the inverse $P^{-1} = (\gamma^4)^{-1} = \gamma^4$:

$$Pu^\alpha(\vec{p}, p_0) = u^\alpha(-\vec{p}, p_0), \qquad \alpha = 1, 2 \quad \text{(particles)}$$
$$Pu^\alpha(\vec{p}, p_0) = -u^\alpha(-\vec{p}, p_0), \qquad \alpha = 3, 4 \quad \text{(anti-particles)}$$

The γ-matrices transform under this transformation as a vector whose first three components change sign:

$$P^{-1}\gamma^\mu P = -\gamma^\mu, \text{ if } \mu = 1,2,3,5, \quad P^{-1}\gamma^4 P = \gamma^4.$$

Charge conjugation not only changes from a particle to an anti-particle, but also changes a u into a $\bar{u}$ and v.v. At the same time γ-matrices are reflected (notation as usual, $\tilde{\gamma}$), thus entirely reversing a string of γ-matrices enclosed between a $\bar{u}$ and a u, which is the type of thing encountered in a diagram. With our choice of γ's the charge conjugation matrix C equals $\gamma^2\gamma^4$, and $C^{-1} = \tilde{C} = -C$.

$$C\bar{u}^1(p) = -u^4(p) \qquad C\bar{u}^2(p) = -u^3(p)$$
$$u^3(p)C^{-1} = \bar{u}^2(p) \qquad u^4(p)C^{-1} = \bar{u}^1(p).$$

Transformation of the γ's:

$$C^{-1}\gamma^\mu C = -\tilde{\gamma}^\mu, \text{ if } \mu = 1,2,3,4, \quad C^{-1}\gamma^5 C = \tilde{\gamma}^5.$$

Time reversal involves exchanging initial and final states as well as changing the direction of spin and three-momentum. This also implies transformation of a u to an $\bar{u}$ and reversal of a string. The associated matrix is $T = -\gamma^5 C$, with C as above. In addition the matrix γ^4 occurs, as time reversal also implies complex conjugation. For example $\tilde{\gamma}^4\bar{u} = \bar{u}\gamma^4 = u^*$.

$$T\tilde{\gamma}^4\bar{u}^1(\vec{p},p_0) = -u^2(-\vec{p},p_0) \qquad T\tilde{\gamma}^4\bar{u}^2(\vec{p},p_0) = u^1(-\vec{p},p_0)$$
$$u^3(\vec{p},p_0)\tilde{\gamma}^4 T^{-1} = \bar{u}^4(-\vec{p},p_0) \qquad u^4(\vec{p},p_0)\tilde{\gamma}^4 T^{-1} = -\bar{u}^3(-\vec{p},p_0).$$

The transformation of the γ's is simply $T^{-1}\gamma^\mu T = \tilde{\gamma}^\mu$ for all μ.

Exercise 4.4 Show that the CPT transformation applied to the anti-particle spinor with spin down gives the particle spinor with spin up. Show that the explicit form of this transformation is γ^5. A CPT transformation is the result of the successive transformations time reversal, space reversal and charge conjugation. This also exchanges initial and final states.

4.3 Fermion Fields

We must now discuss Hilbert space in some more detail. As noted before, the number of states increases considerably, in particular one-particle states by a factor of 4, two-particle states by a factor of 16, etc. In principle we should discuss the precise way in which these states transform under Lorentz transformation, but in practice we do not need that. The important thing is how fields transform, and we therefore concentrate on that.

We will simply define fields, and that is really quite straightforward. In principle we have four fields, but we take them together in one four-component field that transforms under Lorentz transformations in the way discussed above. For example, concentrating on the term in the fields for a particle at rest then we want for the case of a particle with spin up only the first component of the field to be non-zero. That is then the field corresponding to this kind of particle. Thus we multiply the annihilation operator for one particle at rest with spin up with the vector $u^1 = (1, 0, 0, 0)$, so that $\psi(x) = u^1 a(p) \exp(ipx)$ has indeed only the first component non-zero. One can only make choices for particles at rest, the others follow by Lorentz transformations. A Lorentz transformation of the fields translates directly into a Lorentz transformation of the u, and one arrives at the construction that we will describe now. The fields are not real, and we must consider side by side the field and its hermitian conjugate.

There are two four-component fields, $\psi_a(x)$ and $\bar{\psi}_a(x)$, with $a = 1, \ldots, 4$. The latter, $\bar{\psi}(x)$, is related to $\psi(x)$:

$$\bar{\psi}(x) = \psi^\dagger(x)\gamma^4$$

The definition in terms of creation and annihilation operators is:

$$\psi(x) = \frac{1}{\sqrt{V}} \sum_{\vec{k}} \{ a^1(\vec{k})u^1(\vec{k})e^{ikx} + a^2(\vec{k})u^2(\vec{k})e^{ikx}$$

$$+ b^{3\dagger}(\vec{k})u^3(\vec{k})e^{-ikx} + b^{4\dagger}(\vec{k})u^4(\vec{k})e^{-ikx} \}.$$

As usual $k_0 = \sqrt{\vec{k}^2 + m^2}$. The creation and annihilation operators in this expression are:

$a^1(\vec{k})$ Annihilation operator for electrons with momentum $\vec{k}$, spin up

$a^2(\vec{k})$ Idem, spin down

$b^{3\dagger}(\vec{k})$ Creation operator for positrons with momentum $\vec{k}$, spin down

$b^{4\dagger}(\vec{k})$ Idem, spin up

Thus, for example,

$$a^1(\vec{k})|\vec{k}, \uparrow> = |0>$$

and

$$b^{3\dagger}(\vec{k})|0> = |\underline{\vec{k}}, \downarrow> .$$

The above also shows our notation: to indicate an anti-particle we underline its momentum. Note that there is no factor $1/\sqrt{k_0}$ in the expression for ψ. One might say that this factor is contained in the spinors u.

There is a problem with the fields ψ and $\bar{\psi}$, which is that they are not local. Indeed, if we use these fields without further modification in the S-matrix then unacceptable physics results. It is quite a lot of work to show the non-local effects, but it is possible to create situations where particles emerge from a collision at a point far removed from the point of collision. That might be considered as something different from a-causality, where for example particles emerge from a collision before the actual collision by the inital particles takes place, but in fact, what appears non-local for one observer may be a-causal to another. To study these effects one must construct localized wave packets, and collide these packets. A complication is that in a relativistic theory it is impossible to construct sharply delimited wave packets. There are always exponential tails. That complicates the discussion.

For simplicity here we will define locality as a mathematical property of the interaction Hamiltonian, namely we want the commutator of $\mathcal{H}(x)$ and $\mathcal{H}(y)$ to be zero if x and y are outside each other's light cone, which means that if you want to go from x to y or vice versa you must go with a speed exceeding that of light. And if ever we want to do a complete job, then, once we know how an interaction Hamiltonian leads to observable results, we should check this point. So far people tend to postpone such work to "next year".

Let us now investigate this point in some more depth. The various commutators can be worked out without difficulty. While $\psi(x)$ and $\bar{\psi}(x)$ commute with themselves, we have

$$[\psi_a(x), \bar{\psi}_b(y)] = \frac{1}{V} \sum_{\vec{k}} \left[\sum_{i=1}^{2} u_a^i(\vec{k})\bar{u}_b^i(\vec{k})e^{ik(x-y)} \right.$$
$$\left. - \sum_{i=3}^{4} u_a^i(\vec{k})\bar{u}_b^i(\vec{k})e^{-ik(x-y)} \right]$$

$$= \sum_{\vec{k}} \frac{1}{2Vk_0} \left\{ (-i\gamma k + m)_{ab} e^{ik(x-y)} \right.$$

$$\left. + (i\gamma k + m)_{ab} e^{-ik(x-y)} \right\}$$

$$= (-\gamma^\mu \frac{\partial}{\partial x_\mu} + m)_{ab}$$

$$\sum_{\vec{k}} \frac{1}{2Vk_0} \left\{ e^{ik(x-y)} + e^{-ik(x-y)} \right\}.$$

There is a sign difference compared with the scalar field case, and for equal times we do not find zero.

The solution to this problem was found by Born and Jordan. The basic idea is to introduce the exclusion principle, which was at that time already a known experimental fact (for example, no more than two electrons spin up and spin down in the lowest energy state of hydrogen). Let us see now what happens to the mathematics of the various operators if we introduce this exclusion principle.

Limiting ourselves to some definite momentum and also only spin up we have only two possible states: vacuum and a state corresponding to one electron (with that particular momentum) with spin up. The operators a and $a^\dagger$ work as usual:

$$a|1> = |0>, \qquad a|0> = 0, \qquad a^\dagger|0> = |1>.$$

Since there is no two-particle state we must take

$$a^\dagger|1> = 0.$$

In this two-dimensional subspace of Hilbert space these two matrices have the form:

$$a = \begin{bmatrix} 0 & 1 \\ 0 & 0 \end{bmatrix} \qquad a^\dagger = \begin{bmatrix} 0 & 0 \\ 1 & 0 \end{bmatrix}.$$

We have that the square of a and $a^\dagger$ is zero:

$$a^2 = a^{\dagger 2} = 0$$

Further:

$$aa^\dagger = \begin{bmatrix} 1 & 0 \\ 0 & 0 \end{bmatrix} \qquad a^\dagger a = \begin{bmatrix} 0 & 0 \\ 0 & 1 \end{bmatrix}.$$

Thus:

$$aa^\dagger + a^\dagger a \equiv \{a, a^\dagger\} = 1.$$

This is called the anti-commutator of a and $a^\dagger$. Since evidently $aa + aa = 0$ (similar for $a^\dagger$) we have:

$$\{a, a\} = \{a^\dagger, a^\dagger\} = 0 \qquad \{a, a^\dagger\} = \{a^\dagger, a\} = 1$$

A very simple rule emerges: where before commutation rules held we now have anticommutation rules. Going from scalar particles (such as π and σ) to electrons we simply replace [ ] by { } and the equations remain the same. To really achieve this we must be careful if we have two or more particles of different momenta. We will illustrate this on the case of two particles, with momenta $\vec{p}$ and $\vec{k}$ respectively. We now have four states (restricting ourselves to spin up):

$$|0>, \quad |\vec{p}>, \quad |\vec{k}>, \quad |\vec{p}\vec{k}>$$

In this four-dimensional subspace of Hilbert space we take:

$$a(\vec{p}) = \begin{bmatrix} 0 & 1 & 0 & 0 \\ 0 & 0 & 0 & 0 \\ 0 & 0 & 0 & 1 \\ 0 & 0 & 0 & 0 \end{bmatrix} \qquad a(\vec{k}) = \begin{bmatrix} 0 & 0 & 1 & 0 \\ 0 & 0 & 0 & -1 \\ 0 & 0 & 0 & 0 \\ 0 & 0 & 0 & 0 \end{bmatrix}$$

and $a^\dagger(\vec{p})$ and $a^\dagger(\vec{k})$ follow by reflection. The set of matrices a and $a^\dagger$ is fully anticommuting.

$$\{a(\vec{p}), a(\vec{p})\} = \{a^\dagger(\vec{p}), a^\dagger(p)\} = 0$$

and similarly for $a(\vec{k})$ and $a^\dagger(k)$. Further

$$\{a(\vec{p}), a^\dagger(\vec{k})\} = 0$$

and so on. The only non-zero anticommutators are:

$$\{a(\vec{p}), a^\dagger(\vec{p})\} = \{a(\vec{k}), a^\dagger(\vec{k})\} = 1$$

Exercise 4.5 Verify these equations. Check that

$$a(\vec{p})|\vec{p}\vec{k}>= |\vec{k}>, \quad \text{and} \quad a^\dagger(\vec{p})|\vec{k}>= |\vec{p}\vec{k}> .$$

As we have more and more particles it becomes more complicated to keep track of the minus signs that must be put to obtain the correct anticommutation rules. There is a simple way to do that, namely by keeping track of the way particles are ordered in the state. We define $|\vec{p}\vec{k}>$ as the state obtained by first applying $a^\dagger(\vec{k})$ and then $a^\dagger(\vec{p})$ to $|0>$. Then we define $|\vec{k}\vec{p}>= -|\vec{p}\vec{k}>$.

Thus applying first $a^\dagger(\vec{p})$ and then $a^\dagger(\vec{k})$ to $|0>$ we must obtain $-|\vec{p}\vec{k}>$. This defines the minus sign in the matrix $a(\vec{k})$ above.

Clearly, if we stick to this rule (exchange of two particles gives a minus sign) then anticommutators will be zero.

Let us go back to the fields ψ and $\bar{\psi}$. Now the a, $a^\dagger$, b and $b^\dagger$ obey anticommutation rules. Going again through the usual kind of calculation we now find for the anticommutation rules of ψ and $\bar{\psi}$ that:

$$\{\psi, \psi\} = 0 \qquad \{\bar{\psi}, \bar{\psi}\} = 0$$

but

$$\{\psi_\alpha(x), \bar{\psi}_\beta(y)\} = (-\gamma^\mu \frac{\partial}{\partial x_\mu} + m)_{\alpha\beta}$$

$$\sum_{\vec{k}} \frac{1}{2Vk_0} \{e^{ik(x-y)} - e^{-ik(x-y)}\}$$

$$= (-\gamma^\mu \partial_\mu + m)_{\alpha\beta} \Delta_c(x-y)$$

with the same Δ_c as before. This Δ_c was a local function (i.e., $\Delta_c = 0$ if x and y outside each other's light cone), and we see that the anticommutator of $\bar{\psi}$ and ψ is local.

The physically important quantities, however, are not the fields, but the interaction Hamiltonians. These determine the outcome of physically measurable processes. The Hamiltonian must be local in the usual sense, that is that the commutator (not the anti-commutator) is zero outside the light cone. What good is it that the fields obey local anticommutation rules?

The important point now is that if the anticommuting fields appear in even numbers as factors in the Hamiltonian then local anticommutation rules of the fields imply local commutation rules for this Hamiltonian. Suppose, for example, that $\psi(x)\bar{\psi}(y) = -\bar{\psi}(y)\psi(x)$ and similarly for other ψ combinations. Suppose:

$$\mathcal{H}(x) = f(x)\bar{\psi}(x)\psi(x)$$

where f(x) is some numerical factor. If x and y are outside each other's lightcone we have:

$$\mathcal{H}(x)\mathcal{H}(y) = f(x)f(y)\bar{\psi}(x)\psi(x)\bar{\psi}(y)\psi(y)$$
$$= -f(x)f(y)\bar{\psi}(x)\bar{\psi}(y)\psi(x)\psi(y)$$
$$= f(x)f(y)\bar{\psi}(y)\bar{\psi}(x)\psi(x)\psi(y)$$

$$= f(x)f(y)\bar{\psi}(y)\psi(y)\bar{\psi}(x)\psi(x)$$
$$= \mathcal{H}(y)\mathcal{H}(x).$$

It follows

$$[\mathcal{H}(x),\mathcal{H}(y)] = 0$$

if x and y are outside each other's light cone, and the interaction Hamiltonian is local in the usual sense.

The total upshot of all this is:

- Fermion fields have spinors as extra factors;
- Instead of Δ_c we now have $(-\gamma^\mu \partial_\mu + m)\Delta_c$;
- Fermion fields must appear in pairs in the interaction.

4.4 The E.M. Field

As we have stated before, we will suppose that the photon has a small mass κ. The discussion is in fact very similar to the fermion case, except that we have the Lorentz transformations themselves instead of the spinorial transformations X. Rather than giving again all the arguments we simply write down the result for the fields. We now have the e.m. field $A_\mu(x)$ with

$$A_\mu(x) = \sum_{\vec{k}} \frac{1}{\sqrt{2Vk_0}} \sum_{i=1}^{3} \left\{ e_\mu^i(k)a^i(\vec{k})e^{ikx} + \bar{e}_\mu^i(k)a^{\dagger i}(k)e^{-ikx} \right\}$$

As usual $k_0 = \sqrt{\vec{k}^2 + \kappa^2}$. In here the $e_\mu^i(k)$ are a set of three vectors obeying the equation

$$k_\mu e_\mu(k) = 0.$$

Consider now a particle of spin 1 at rest. Such a particle has three possible polarization states, and this can be represented by three basis vectors in a four-dimensional space:

$$e_\mu^1 = \begin{pmatrix} 1 \\ 0 \\ 0 \\ 0 \end{pmatrix} \quad e_\mu^2 = \begin{pmatrix} 0 \\ 1 \\ 0 \\ 0 \end{pmatrix} \quad e_\mu^3 = \begin{pmatrix} 0 \\ 0 \\ 1 \\ 0 \end{pmatrix}$$

We must work in a four-dimensional space because there is no representation of the Lorentz group in three dimensions. In the rest system the four-momentum has the form $k = (0,0,0,i\kappa)$, and these three vectors can be seen as the solutions of the equation $k_\mu e_\mu = 0$.

The vector $\bar{e}_\mu$ is the complex conjugate of e_μ, except that the i of the fourth component is not conjugated. If we stick to the prescription of not conjugating this Lorentz i, then Lorentz transformations are real transformations, and if e_μ transforms as a vector then also $\bar{e}_\mu$ transforms as a vector. Thus, if $e'_\mu = L_{\mu\nu} e_\nu$ then

$$\bar{e}'_\mu = \bar{L}_{\mu\nu} \bar{e}_\nu = L_{\mu\nu} \bar{e}_\nu .$$

Usually one takes for the e_μ real vectors (in the Lorentz sense), as shown above, and then there is no difference between e and $\bar{e}$. The polarization vectors shown above are called the linear polarization vectors. The choice $e^\pm = (e^1 \pm ie^2)/\sqrt{2}$ instead of e^1 and e^2 will be discussed elsewhere.

Our normalization is that $e_\mu \bar{e}_\mu = 1$. The previous discussion shows that this is Lorentz invariant, as should be.

For the e^i_μ the following equation holds:

$$\sum_i e^i_\mu(k)\bar{e}^i_\nu(k) = \sum_i \bar{e}^i_\mu(k)e^i_\nu(k) = \delta_{\mu\nu} + \frac{k_\mu k_\nu}{\kappa^2}$$

Exercise 4.6 Verify this. Hint: Go first to the rest system.

Using this, one derives the commutator:

$$[A_\mu(x), A_\nu(y)] = \left(\delta_{\mu\nu} - \frac{1}{\kappa^2}\partial_\mu\partial_\nu\right)\Delta_c(x-y).$$

Obviously this commutator is local, because Δ_c is local.

Under Lorentz transformations the field A_μ transforms as a vector:

$$A'_\mu(x') = L_{\mu\nu}A_\nu(x), \quad x' = Lx.$$

The advantage of working with a photon with a rest mass is that we can go to the rest frame and restrict ourselves to the three real vectors e_μ. Even if the relativistic i will give minus signs here and there, Lorentz invariance of dot-products assures us that what is positive in one frame will remain so in any other. If we would use also the vector $(0,0,0,i)$ then trouble arises, and one may find negative "probabilities". This problem is always very near in theory, whether through the γ^4 of spinors, or the imaginary component of a four-vector.

4.5 Quantum Electrodynamics

As before we now develop the formalism by introducing equations of motion, interaction Hamiltonian and S-matrix. The "in" fields obey the free field equations:

$$(\Box - \kappa^2)A_\mu(x) = 0$$

$$\partial_\mu A_\mu(x) = 0$$

$$(\gamma^\mu \partial_\mu + m)\psi(x) = 0$$

The second equation is called the **Lorentz condition**. It follows because of the equation $k_\mu e_\mu(k) = 0$.

The simplest interaction such that Lorentz invariance is maintained is:

$$(\Box - \kappa^2)A_\mu(x) = ie(\bar{\psi}(x)\gamma^\mu \psi(x)).$$

Indeed, under a Lorentz transformation:

$$\psi \to X\psi$$

$$\bar{\psi} \to \bar{\psi}X^{-1}.$$

Consequently:

$$(\bar{\psi}\gamma^\mu \psi) \to (\bar{\psi}X^{-1}\gamma^\mu X\psi)$$

$$= L_{\mu\nu}(\bar{\psi}\gamma^\nu \psi).$$

Thus the right hand side of the above equation of motion transforms as a vector, just as the left hand side. The i has been introduced to make the right hand side hermitian:

$$(\bar{\psi}\gamma^\mu \psi)^\dagger = (\psi^\dagger \gamma^4 \gamma^\mu \psi)^\dagger$$

$$= (\psi^\dagger \gamma^\mu \gamma^4 \psi).$$

We used $(ABCD)^\dagger = D^\dagger C^\dagger B^\dagger A^\dagger$. Now $\gamma^\mu \gamma^4 = -\gamma^4 \gamma^\mu$ if $u = 1, 2, 3$ and we see that the first three components of $(\bar{\psi}\gamma_\mu \psi)$ are imaginary (antihermitian). The i in front compensates this. The fourth component gets no minus sign, is real, becomes imaginary after multiplication with i. This is of course precisely what we want since also the fourth component of A_μ is imaginary.

This equation of motion derives from the interaction Hamiltonian:

$$\mathcal{H}(x) = ieA_\mu(x)(\bar{\psi}(x)\gamma^\mu \psi(x)).$$

The equation of motion for the ψ field consistent with this is:

$$(\gamma^\mu \partial_\mu + m)\psi(x) = ieA_\mu \gamma^\mu \psi.$$

In deriving the S-matrix no substantial difficulties arise. Mainly we get extra factors $u(\vec{k})$ and $e_\mu(k)$, as well as

$$(\gamma^\mu \partial_\mu + m) \quad \text{or} \quad (-i\gamma^\mu k_\mu + m)$$

and

$$\delta_{\mu\nu} - \partial_\mu \partial_\nu / \kappa^2 \quad \text{or} \quad \delta_{\mu\nu} + k_\mu k_\nu / \kappa^2.$$

There are some slight technical difficulties relating to derivatives of the θ functions in the time-ordered products. That will not be discussed in this chapter. We straight away give a table of the Feynman rules of this theory.

Incoming electron	$V^{-1/2} u^\lambda(\vec{k})\quad \lambda = 1,2$	
Incoming positron	$-V^{-1/2} \bar{u}^\lambda(\vec{k})\quad \lambda = 3,4$	
Outgoing electron	$V^{-1/2} \bar{u}^\lambda(\vec{k})\quad \lambda = 1,2$	
Outgoing positron	$V^{-1/2} u^\lambda(\vec{k})\quad \lambda = 3,4$	
Propagator:	$\dfrac{-i\gamma^\mu k_\mu + m}{k^2 + m^2 - i\epsilon}\quad$ k in direction of the arrow	
Incoming photon	$\dfrac{1}{\sqrt{2V k_0}} e^i_\mu(k)\quad$ i $= 1,2,3$	
Outgoing photon	$\dfrac{1}{\sqrt{2V k_0}} \bar{e}^i_\mu(k)\quad$ i $= 1,2,3$	
Propagator:	$\dfrac{\delta_{\mu\nu} + k_\mu k_\nu / \kappa^2}{k^2 + \kappa^2 - i\epsilon}$	
Vertex:	$-ie\gamma^\mu$	

In addition we have the following factors:

   i $(2\pi)^4 i$ for any vertex
   ii $-i(2\pi)^{-4}$ for any propagator
   iii $(-1)^\ell$ for any diagram containing fermion lines
   iv   A combinatorial factor (see appendix on combinatorial factors)
   v -1 for every closed loop of fermions

A factor -1 for an incoming anti-fermion is included above.

Spinor and γ matrix ordering. In problems with fermion lines there is the matter of ordering the spinors and γ matrices. Now note that $\bar{u}$ is in $\bar{\psi}$, always occurring on the left. This gives the rule: for a given fermion line start at the end of that line and read **against** the arrow. Write down the factors correspondingly. For a closed loop start at any vertex on the line and read against the arrow until back at the original vertex.

This means that non-loop lines always start with a $\bar{u}$ and finish with a u, with inbetween the various factors encountered while reading, such as propagator numerators $-i\gamma p + m$ and vertex factors γ^{μ}. Closed loop lines give directly a trace.

Ad iii): Because of the Fermi–Dirac statistics any interchange of two fermions (internal or external) changes the sign of the contribution in the integrand; ℓ has to be chosen so that a diagram obtained from another diagram by permutation of electron lines has the correct sign relative to that diagram. Thus in a given order the relative signs of diagrams are determined by this recipe. To obtain the relative sign of diagrams of different order one must give equal sign to diagrams that can be made equal by taking away vertices. Examples of diagrams having different sign, that is ℓ differing by one, are shown in the figures. The relative minus sign of the first two diagrams is formulated explicitly by the prescription that there is a minus sign for every incoming anti-fermion, here incoming positron.

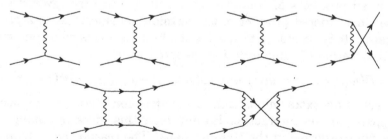

The next figure shows two diagrams having equal ℓ, the first follows from the second by taking away a photon line.

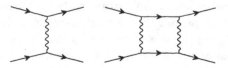

Ad iv): The combinatorial factor is one over the number of diagrams that can be obtained by interchange of lines such that the diagram remains topologically the same. For diagrams of low order this factor can easily be found; in quantum electrodynamics it is always one. In the $\sigma\pi\pi$ theory these factors are non-trivial. The first two diagrams shown in the figure are the same, assuming that the internal lines refer to the same boson.

Here we have therefore a factor 1/2. Somewhat curiously, also the tadpole diagram shown must be given a factor 1/2, which requires a twisted mind (twisting the circle to an horizontal eight) to understand. See appendix.

Ad v): Consider a very simple diagram, see figure. The corresponding expression arises, as far as the fermion part is concerned, from the product of four fields that results in two propagators:

$$P\{\bar{\psi}(x)\psi(x)\bar{\psi}(x')\psi(x')\} = -P\{\bar{\psi}(x)\psi(x')\} \cdot P\{\bar{\psi}(x')\psi(x)\}.$$

There is always such a minus sign for a closed loop.

There are some technical details in the derivation of the Feynman rules relating to the fermion fields. One encounters the time ordered products of Hamiltonians, and this in turn gives rise to the time ordered product of the fermion fields leading to a propagator. It is necessary to use for the fermion fields what one may call a "fermion" time ordering, denoted by P:

$$P(\bar{\psi}(x)\psi(y)) = \theta(x_0 - y_0)\bar{\psi}(x)\psi(y) - \theta(y_0 - x_0)\psi(y)\bar{\psi}(x)$$

There is an extra minus sign. Since fermion fields always occur in pairs in the interaction Hamiltonian this gives no change in the time ordering of the Hamiltonians. The reason that one must do this is that one wants the two terms in such a time ordered product to become equal for equal times ($x_0 = y_0$). Else we would have singular behaviour going from $x_0 > y_0$ through $x_0 = y_0$ to $x_0 < y_0$, and the propagator would become quite horrible. Of course, it would combine with other factors to be all right in the end, but the use of the P-product avoids such problems.

4.6 Charged Vector Boson Fields

The vector bosons of weak interactions have been observed and are part of the Standard Model. They are described by fields much like the photon field, except that the photon is its own anti-particle, while the anti-particle of the W^- is the W^+. Like with the fermions, the creation operator for the anti-particle is combined with the annihilation operator of the particle. Thus:

$$W_\mu^-(x) = \sum_{\vec{k}} \frac{1}{\sqrt{2Vk_0}} \sum_{i=1}^{2} \left\{ e_\mu^i(k) a^i(\vec{k}) e^{ikx} + \bar{e}_\mu^i(k) b^{\dagger i}(k) e^{-ikx} \right\}.$$

where $b^{\dagger i}(k)$ is the creation operator for a positively charged vector boson with polarization as described by the polarization vector $\bar{e}_\mu^i$. Further, $W_\mu^+ = (W_\mu^-)^\dagger$. This set of fields is local, they all commute outside the light cone. That is in fact the reason why the above combinations are taken.

It is also possible to keep on working with real fields. Writing

$$W^- = \frac{1}{\sqrt{2}}(W^1 + iW^2) \quad \text{and} \quad W^+ = \frac{1}{\sqrt{2}}(W^1 - iW^2),$$

one has real fields W^1 and W^2, satisfying locality properties. The resulting theory gives the same results. Which formulation to use is a matter of convenience.

In the charged field formulation one has a propagator that must be given an arrow to indicate the sign of the charge flowing through the line. Such an oriented propagator corresponds to the time-ordered product

$$< 0 \mid T \left\{ W_\mu^-(x) W_\nu^+(y) \right\} \mid 0 >.$$

We will very briefly outline how this works on a simple example.

Consider a theory involving a scalar neutral particle σ coupled to a charged vector boson W. The interaction Hamiltonian is:

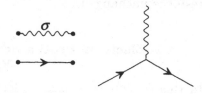

$$\mathcal{H}(x) = g\sigma(x) W_\mu^-(x) W_\mu^+(x).$$

The vertices and propagators of this theory are shown in the figure.

In a vertex one may annihilate a W^- (the W^- field) and emit a W^- (in the W^+ field). Or annihilate a W^+ and emit a W^+ etc. The arrow in the propagator has no implications with respect to the expression to be

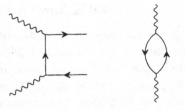

used for that propagator (which is precisely the same as for a massive photon), because that expression has no terms linear in the momentum. Its function is to keep track of the flow of charge. The first diagram shows annihilation of two σ into a W^- and a W^+. There is only one diagram. The next shows a σ self-energy diagram. The combinatorial factor is one, there is no reflection symmetry because of the arrows.

It is interesting to consider this self-energy diagram in the real field formulation. The interaction Hamiltonian is now:

$$\mathcal{H}(x) = \frac{g}{2}\sigma W^1 W^1 + \frac{g}{2}\sigma W^2 W^2 \,.$$

There are two vector fields, indicated by a straight and a dashed line. The vertices and propagators are as shown in the figure. Note that the factor $1/2$ disappears in the Feynman rules for the vertices as one has two identical fields in each vertex. There are now two σ self-energy diagrams, but they have each a combinatorial factor $1/2$, and the net result is unchanged.

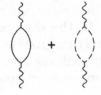

4.7 Electron–Proton Scattering. The Rutherford Formula

In this section we consider electron–proton scattering. This example is very important, because we know what in the non-relativistic limit the result must be: the Rutherford formula for scattering of electrons in a Coulomb field. This example thus serves as a test whereby we can verify that we have chosen the correct equations of motion, with the correct interaction Hamiltonian. In lowest non-vanishing order we have only the diagram shown, if we assume the same type of interaction Hamiltonian for proton–photon interaction:

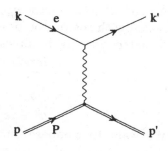

$$\mathcal{H}(x) = -ieA_\mu(x)\{\bar{\psi}_e(x)\gamma^\mu\psi_e(x)\} + ieA_\mu(x)\{\bar{\psi}_P(x)\gamma^\mu\psi_P(x)\}$$

The charge of the proton is opposite to that of the electron. The electron field is denoted by ψ_e and the proton field by ψ_P. Only one type of interchange is allowed: $x_1 \leftrightarrow x_2$. Because the particles are not identical we cannot exchange k and p without exchanging p' and k'. Thus there is a factor 2, which cancels agains the 1/2! in the S-matrix element. As in the previous chapter we derive

$$< k'p'|S|kp > = \frac{(-i)^2}{2} \int d_4x_1 \, d_4x_2$$

$$< \vec{k}'\vec{p}''|P\{\mathcal{H}(x_1)\mathcal{H}(x_2)\}|\vec{k},\vec{p} >$$

$$= \frac{(i)^2 e^2 i(-i)(2\pi)^8}{V^2 (2\pi)^4 i} \int d_4q \, \delta_4(q+k-k')$$

$$\delta_4(q-p+p')$$

$$\cdot \frac{\delta_{\mu\nu}}{q^2+\kappa^2-i\epsilon} \cdot \{\bar{u}(\vec{k}')\gamma^\mu u(\vec{k})\} \, \{\bar{u}(\vec{p}')\gamma^\nu u(\vec{p})\}$$

$$= \frac{e^2 i(2\pi)^4}{V^2} \delta_4(p+k-p'-k')$$

$$\cdot \frac{\delta_{\mu\nu}}{q^2+\kappa^2-i\epsilon} \{\bar{u}(\vec{k}')\gamma^\mu u(\vec{k})\}\{\bar{u}(\vec{p}')\gamma^\nu u(\vec{p})\}$$

where $q = k' - k = p - p'$. The factor $(1/\kappa^2)q_\mu q_\nu$ in the photon propagator disappears, for if we consider the product

$$q_\mu\{\bar{u}(\vec{k}')\gamma^\mu u(k)\} = \bar{u}(\vec{k}')\{\gamma^\mu k'_\mu - \gamma^\mu k_\mu\}u(\vec{k}) = \ldots$$

we can apply the Dirac equation $(i\gamma k + m)u = 0$ and similarly for $\bar{u}$, which gives

$$\ldots = \bar{u}(\vec{k}')\{im - im\}u(\vec{k}) = 0.$$

Apparently, the factor $q_\mu q_\nu$ gives no contribution after multiplication with the proton or electron current. In the limit $\kappa \to 0$ the S-matrix element becomes

$$<S> = \frac{ie^2(2\pi)^4}{V^2} \delta_4(p + k - p' - k')$$

$$\frac{\delta_{\mu\nu}}{q^2 - i\epsilon} \{\bar{u}(\vec{k}')\gamma^\mu u(\vec{k})\}\{\bar{u}(\vec{p}')\gamma^\nu u(\vec{p})\}.$$

We want to compare our results with the classical Rutherford cross section, and so we must calculate the total scattering cross section σ_{tot}. This σ_{tot} equals the transition probability $|<S>|^2$ integrated over all momentum space of the outgoing particles. The number of states within a momentum space element d_3q is $V/(2\pi)^3 \cdot d_3q$, for periodic boundary conditions over a volume V. Furthermore, the incoming particles must be normalized to unit flux, that is one particle per unit surface per unit time. For one particle in V the flux is v/V, where v is the velocity $|\vec{k}|/k_0$.* We have:

$$\sigma_{tot} = \int d_3k'\, d_3p' \frac{V^2}{(2\pi)^6} \frac{Vk_0}{|\vec{k}|} \sum_{\text{spins}} |<S>|^2$$

where the sum extends over all spins of the outgoing particles. The only problem in evaluating $|<S>|^2$ is the behaviour of the δ-function. As shown earlier, squaring the δ-function gives a factor $VT/(2\pi)^4$, and taking the transition probability per unit time implies dividing out T. The squared δ_4-function then becomes $V/(2\pi)^4\delta_4(k + p - k' - p')$, and substitution into the expression for σ_{tot} gives finally

$$\sigma_{tot} = \int d_3p'\, d_3k' \frac{e^4}{(2\pi)^2} \frac{k_0}{|\vec{k}|} \delta_4(k + p - k' - p') \frac{1}{q^4}$$

$$\sum_{\text{spins}} |\{\bar{u}(\vec{k}')\gamma^\mu u(\vec{k})\}\{\bar{u}(\vec{p}')\gamma^\mu u(\vec{p})\}|^2$$

where we performed summation over ν so that only μ is left. For further evaluation we consider the last factor of the integrand.

* Instead of $1/v$, for beam collisions in a storage ring the flux factor is

$$\sqrt{\frac{p_{10}^2 p_{20}^2}{(p_1 p_2)^2 - m_1^2 m_2^2}}$$

where p_1 and p_2 are the momenta of the incoming particles.

We must now work out:
$$(\bar{u}(\vec{k}')\gamma^\mu u(\vec{k}))\,(\bar{u}(\vec{k}')\gamma^\nu u(\vec{k}))^*$$
multiplied by an analogous expression for the proton spinors. We write:
$$|\bar{u}(\vec{k}')\gamma^\mu u(\vec{k})|^2 = |\bar{u}'_i\,\gamma^\mu_{ij}\,u_j|^2 = \ldots.$$
Substituting the definition of $\bar{u}$, we get
$$\ldots = |u'^*_k\,\gamma^4_{ki}\gamma^\mu_{ij}u_j|^2 = u'^*_k\,\gamma^4_{ki}\gamma^\mu_{ij}\,u_j\cdot u'_a\gamma^{4*}_{ab}\gamma^{\nu*}_{bc}\,u^*_c = \ldots$$
where summation over the values 1–4 of the indices k, i, j, a, b and c is understood. If we remember $\gamma^{\mu*} = \tilde{\gamma}^\mu$, and that $\gamma^\mu\gamma^4 = -\gamma^4\gamma^\mu$ if $\mu \neq 4$, we obtain
$$\ldots = (\bar{u}'\gamma^\mu u)\,(u'_a\gamma^4_{ba}\gamma^\nu_{cb}u^*_c) = (\bar{u}'\gamma^\mu u)(u^*_c\gamma^\nu_{cb}\,\gamma^4_{ba}\,u'_a)$$
$$= -\,(\bar{u}'\,\gamma^\mu u)\,(\bar{u}\gamma^\nu u')\,(-1)^{\delta_{\nu 4}}.$$
The exponent $\delta_{\nu 4}$ appears because of the anticommutation rules for γ-matrices; in other words, $\delta_{\nu 4}$ occurs because in the relativistic formulation the fourth component of a vector is always purely imaginary, and thus changes sign upon complex conjugation.

Next, we must evaluate the sum over spins of the outgoing particles. If we assume that the initial polarizations of the particles are not known, we must also take the average over the incoming spins, which amounts to summation and division by the possible number of configurations. These summations are carried out by means of the formula
$$\sum_{i=1}^{2}\bar{u}^i_a(k)u^i_b(k) = -\frac{1}{2k_0}\,(i\gamma k - m)_{ba}$$
which we already encountered before. The sum respectively average over spins then becomes
$$-\tfrac{1}{2}(-1)^{\delta_{\nu 4}}\sum_{i=1}^{2}\sum_{j=1}^{2}(\bar{u}'^i_a\gamma^\mu_{ab}\,u^j_b)\,(\bar{u}^j_c\,\gamma^\nu_{cd}\,u'^i_d)$$
$$= -(1)^{\delta_{\nu 4}}\left[\gamma^\mu_{ab}(-i\gamma k + m)_{bc}\,\gamma^\nu_{cd}\,(-i\gamma k' + m)_{da}\right]\frac{1}{8k_0 k'_0}$$
The first and the last summation index of the expression are equal, which means that the expression is the trace of the matrix between brackets
$$\ldots = \frac{-(-1)^{\delta_{\nu 4}}}{8k_0 k'_0}\,\mathrm{Tr}\left[\gamma^\mu(-i\gamma k + m)\gamma^\nu\,(-i\gamma k' + m)\right].$$

Using the formulae for the trace of products of γ-matrices (see appendix) we obtain finally the sum over electron and proton spin states (from here on M = proton mass):

$$|\bar{u}(\vec{k}')\gamma^\mu u(\vec{k})|^2 \rightarrow -\frac{(-1)^{\delta_{\nu 4}}}{2k_0 k_0'}\{-k_\mu k_\nu' - k_\nu k_\mu' + \delta_{\mu\nu}(kk') + m^2\delta_{\mu\nu}\}$$

$$|\bar{u}(\vec{p}')\gamma^\mu u(\vec{p})|^2 \rightarrow -\frac{(-1)^{\delta_{\nu 4}}}{2p_0 p_0'}\{-p_\mu p_\nu' - p_\nu p_\mu' + \delta_{\mu\nu}(pp') + M^2\delta_{\mu\nu}\}$$

With these expressions, σ_{tot} can be found. For comparison with known results, we make some additional approximations. First, we take our coordinates in the proton rest system. Second, we take the proton very heavy with respect to the electron, i.e., $M \gg m$, and furthermore we suppose the non-relativistic case $m \gg |\vec{k}|$ and $m \gg |\vec{k}'|$. (The comparison of quantities m and $|\vec{k}|$ can be made, because in our system of units ($\hbar = c = 1$) every quantity is expressed in the same unit, namely in MeV.) In the proton rest system we have $p = (0,0,0,iM)$ and $p' = p + k - k' = p - q$. Thus $\vec{p}' = \vec{q}$ and $p'^2 = -M^2$ gives $p_0' = \sqrt{M^2 + \vec{p}'^2} = M + \vec{q}^2/2M +$ $.. = M$ up to **second** order in $|\vec{q}|$. The proton part becomes:

$$\frac{-(-1)^{\delta_{\nu 4}}}{2p_0 p_0'}\{2M^2\delta_{\mu 4}\delta_{\nu 4} - \delta_{\mu\nu}M^2 + \delta_{\mu\nu}M^2\} = \frac{-(-1)^{\delta_{\nu 4}}}{2p_0 p_0'}2M^2\delta_{\mu 4}\delta_{\nu 4}.$$

Multiplication with the electron spinor part gives

$$\frac{2M^2}{4k_0 p_0 k_0' p_0'}\{-2k_4 k_4' + (kk') + m^2\} = \ldots$$

The momenta k' and k together make up the momentum q which is exchanged along the photon line: $k' = k + q$. Furthermore, we have $k^2 = -m^2$ and, in the non-relativistic limit, $k_4 \approx im$ and $|\vec{k}| \ll m$, $|\vec{k}'| \ll m$ so that we obtain

$$\ldots = \frac{M^2\,m^2}{p_0 k_0 p_0' k_0'} = \frac{M^2\,m^2}{Mm\,Mm} = 1.$$

We have now calculated the sum over spins, in the non-relativistic approximation and with infinite proton mass, and the result is simply 1, independent of momenta. This simple result means that spin effects do not play any role in non-relativistic scattering on a heavy particle. The total cross section is

$$\sigma_{\text{tot}} = \int d_3 p'\, d_3 k'\, \frac{e^4 k_0}{(2\pi)^2 |\vec{k}| q^4}\, \delta_4(k + p - k' - p') = \ldots$$

Integrating over $\vec{p}\,'$, substituting polar coordinates for $\vec{k}'$, and using that $p_0 = p_0' + O(\vec{q}^{\,2})$ we obtain:

$$\ldots = 4\alpha^2 \int |\vec{k}'|^2 d|\vec{k}'| d\Omega \, \frac{k_0}{|\vec{k}|} \frac{1}{q^4} \, \delta(k_0 - k_0'),$$

where Ω is a solid angle ($d\Omega = \sin\theta d\theta d\phi$) and α is the fine structure constant $\alpha = e^2/4\pi \simeq 1/137$. The δ-function forces $k_0 = k_0'$, and consequently $|\vec{k}| = |\vec{k}'|$. We call this elastic scattering, the magnitude of the momentum of the electron is unchanged after scattering. Further:

$$k_0' = \sqrt{|\vec{k}'|^2 + m^2} \quad \rightarrow \quad \frac{dk_0'}{d|\vec{k}'|} = \frac{|\vec{k}'|}{\sqrt{|\vec{k}'|^2 + m^2}} = \frac{|\vec{k}'|}{k_0'} \, ,$$

so that $|\vec{k}'|d|\vec{k}'| = k_0 dk_0$, which gives after integration over k_0:

$$\sigma_{\text{tot}} = 4\alpha^2 \int \frac{k_0^2}{q^4} \, d\Omega.$$

With $q = k' - k$, we have

$$q^2 = k^2 + k'^2 - 2(kk') = -2m^2 - 2|\vec{k}| \, |\vec{k}'| \cos\theta + 2k_0 k_0',$$

where θ is the scattering angle between $\vec{k}$ and $\vec{k}'$. Substitution of this and $k_0' = \sqrt{|\vec{k}'| + m^2}$, $k_0 = \sqrt{|\vec{k}|^2 + m^2}$ gives (remember, $|\vec{k}| = |\vec{k}'|$):

$$q^2 = 2|\vec{k}|^2 (1 - \cos\theta) = 4|\vec{k}^2| \sin^2 \frac{\theta}{2}.$$

In the non-relativistic limit $k_0 \simeq m$ and $|\vec{k}| \simeq mv$, where v is the velocity of the incident particle. We finally find

$$\sigma_{\text{tot}} = \frac{\alpha^2}{4} \int \frac{d\Omega}{m^2 v^4 \sin^4(\theta/2)} = \frac{\alpha^2 \pi}{2} \int \frac{\sin\theta \, d\theta}{m^2 v^4 \sin^4(\theta/2)} \, .$$

This is the Rutherford scattering formula. We conclude that in the given approximation the theory agrees with the classical theory and with experiments, which suggests that our choice for the interaction Hamiltonian has been the right one, at least in the approximation involved. This equation is identical to the equation obtained for the spinless case with σ exchange if we substitute $g^2 = e^2 M M_p$ and $M = m$.

5

Explorations

5.1 Scattering Cross Section for $e^+ e^- \to \mu^+ \mu^-$

This is the process observed in electron-positron colliders such as at SLAC in Stanford, or at DESY in Hamburg.

The interaction of muons with photons is assumed to be identical to the electron–photon interaction. In lowest order there is only one diagram. The S-matrix element in lowest order, to be denoted by $\mathcal{M}$, is:

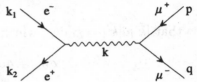

$$< pq|S|k_1 k_2 > = \mathcal{M}\delta_4(k_1 + k_2 - p - q)$$

$$\mathcal{M} = \frac{(2\pi)^4 i}{V^2} (-e^2)(-1) \{\bar{u}(q)\gamma^\mu u(p)\} \frac{1}{k^2} \{\bar{u}(k_2)\,\gamma^\mu u(k_1)\}$$

where e is the electric charge. We have dropped the $k_\mu k_\nu / \kappa^2$ term in the photon propagation (since it gives zero), and also set the photon mass κ to zero in that propagator. A minus sign for the incoming anti-fermion is included.

We now must square this matrix element, average over electron spins and sum over muon spins:

$$\mathcal{M}_s \equiv \frac{1}{4}\sum |\mathcal{M}|^2 = \frac{(2\pi)^8 e^4}{4V^4} \frac{1}{k^4} \frac{1}{2^4 q_0 p_0 k_{10} k_{20}} \cdot$$
$$\mathrm{Tr}\left[(-i\gamma q + M)\gamma^\mu(-i\gamma p - M)\gamma^\nu\right]$$
$$\mathrm{Tr}\left[(-i\gamma k_2 - m)\gamma^\mu(-i\gamma k_1 + m)\gamma^\nu\right]$$

The $1/4$ is for the spin averaging. The muon mass is denoted by M, the electron mass by m. We inserted already the expressions for products of spinors summed over spin states.

The traces may be worked out without any problems, we get:

$$\mathcal{M}_s = \frac{(2\pi)^8 e^4}{16V^4 k^4 q_0 p_0 k_{10} k_{20}} \cdot \frac{16}{4}$$
$$\cdot [-q_\mu p_\nu - q_\nu p_\mu + (pq)\delta_{\mu\nu} - M^2 \delta_{\mu\nu}]$$
$$\cdot [-k_{2\mu} k_{1\nu} - k_{2\nu} k_{1\mu} + (k_1 k_2)\delta_{\mu\nu} - m^2 \delta_{\mu\nu}]$$
$$= \frac{(2\pi)^8 e^4}{16V^4 k^4 q_0 p_0 k_{10} k_{20}} \cdot \frac{16}{4}$$
$$\cdot [2(qk_2)(pk_1) + 2(qk_1)(pk_2) - 2(k_1 k_2)(pq) + 2m^2(pq)$$
$$+ 2(pq)(k_1 k_2) - 4m^2(pq) - 2M^2(k_1 k_2) + 4m^2 M^2]$$
$$= \frac{(2\pi)^8 e^4}{4V^4 k^4 q_0 p_0 k_{10} k_{20}}$$
$$\cdot [2(qk_2)(pk_1) + 2(qk_1)(pk_2) - 2m^2(pq)$$
$$- 2M^2(k_1 k_2) + 4m^2 M^2].$$

We now specialize to a case such as at SLAC or DESY. There the energy of the e^- (and e^+) is generally larger than 1 GeV (= 1000 MeV) and with respect to this the electron mass (≈ 0.5 MeV) and also the muon mass (≈ 100 MeV) may be neglected. Note that only masses squared appear. In observable quantities one never will have a dependence on the sign of a mass.

Next we take the coordinate system as applicable to the above machines, that is the system where e^+ and e^- have the same energy but opposite momentum. After the collision the muons will 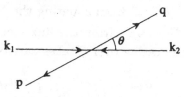 emerge with the same energy, flying off in opposite directions. In this system we have:

$$k_1 = (0, 0, E, iE), \qquad k_2 = (0, 0, -E, iE),$$

where E is now the energy of each electron. Ignoring the electron mass this means for example that $|\vec{k}_1| = \sqrt{k_0^2 - m^2} \simeq k_0 = E$. Note that we take the z-axis along the e^+e^- beam directions. As is clear from the figure there is really only one parameter, namely the angle θ.

Since the muons have the same energy as the electrons we have

$|\vec{p}| = |\vec{q}| = p_0 = q_0 = E$. The values of the various quantities can now be established easily:

$$(pk_2) = (\vec{p}\vec{k}_2) - E^2 = -E^2(1 - \cos\theta) \qquad (pk_1) = -E^2(1 + \cos\theta)$$
$$(qk_1) = -E^2(1 - \cos\theta) \qquad (qk_2) = -E^2(1 + \cos\theta)$$
$$(k_1 k_2) = -2E^2 \qquad (pq) = -2E^2$$
$$k^2 = (k_1 + k_2)^2 = -4E^2 \qquad q_0 = p_0 = k_{10} = k_{20} = E$$

With this we find:

$$|\mathcal{M}|^2 = \frac{(2\pi)^8 e^4}{4V^4 16E^8}\left[2E^4(1 - \cos\theta)^2 + 2E^4(1 + \cos\theta)^2\right]$$

$$= \frac{4(2\pi)^8 e^4}{4V^4 16E^4}(1 + \cos^2\theta) = \frac{(2\pi)^8 e^4}{16V^4 E^4}(1 + \cos^2\theta)$$

The total cross section is:

$$\sigma_{tot} = \int d_3 p \, d_3 q \, \frac{V^4}{(2\pi)^{10}} \frac{E^2}{\sqrt{4E^4}}$$
$$\cdot \frac{(2\pi)^8 e^4}{16V^4 E^4}(1 + \cos^2\theta)\delta_4(k_1 + k_2 - p - q).$$

The various factors are:
− $V/(2\pi)^3$ for each final momentum integration
− $V/(2\pi)^4$ from squaring the δ-function
− $E^2/\sqrt{4E^4}$ from the flux factor $k_{10}k_{20}/\sqrt{(k_1 k_2)^2 - m_1^2 m_2^2}$
Integrating over $\vec{q}$ takes care of three δ-functions:

$$\sigma_{tot} = \frac{e^4}{(2\pi)^2\, 32E^4} \int d_3 p \,(1 + \cos^2\theta)\, \delta(2E - 2p_0)$$

$$= \frac{e^4}{(2\pi)^2\, 32E^4} \int_0^{2\pi} d\phi \int_{-1}^{1} d\cos\theta \,(1 + \cos^2\theta)\frac{E^2}{2}$$

where we did the $|\vec{p}|$ integral. Note that $|\vec{p}| = p_0$, thus $p^2 dp = p_0^2 dp_0 = E^2 dp_0$. There is a factor $1/2$ from the inverse of the derivative of the argument of the δ-function with respect to p_0.

Also the integrations over ϕ and θ can be done:

$$\sigma_{tot} = \frac{e^4}{(2\pi)^2\, 32E^4} \cdot 2\pi \frac{E^2}{2} \left(2 + \frac{2}{3}\right)$$

$$= \frac{e^4}{(2\pi)\, 32E^4} \cdot E^2 \frac{4}{3} = \frac{e^4}{48\pi E^2} = \frac{\alpha^2 \pi}{3E^2}$$

with $\alpha = e^2/4\pi$. Often one uses $s = 2E$ = total energy of the system. Expressing E in MeV we may convert to cm^2 using $\hbar c = 1.973 \times 10^{-11}$ MeV $\cdot$ cm:

$$\sigma_{\text{tot}} = \frac{4\alpha^2 \pi}{3s^2} (1.973)^2 \cdot 10^{-22} \text{ cm}^2.$$

Often the energy is expressed in GeV ($= 10^3$ MeV) and the cross section in nanobarn (1 barn $= 10^{-24}$ cm^2, 1 nb $= 10^{-9}$b $= 10^{-33}$ cm^2):

$$\sigma_{\text{tot}} = \frac{4\alpha^2 \pi}{3s^2} (1.973)^2 \cdot 10^{-28} \cdot 10^{33} \text{ nb}, \ s \text{ in GeV}$$

$$= \frac{86.8}{s^2} \text{ nb } (s \text{ in GeV}).$$

This equation agrees very well with the experimental data, measured to about $s = 100$ GeV.

5.2 Pion Decay. Two Body Phase Space. Cabibbo Angle

Let us next consider π decay. This decay proceeds through weak interactions, and here we will just assume an interaction Hamiltonian for the decay $\pi \to e\bar{\nu}$:

$$\mathcal{H}_i = (\bar{\psi}_e \ \gamma^\alpha(a + b\gamma^5)\psi_\nu) \ \partial_\alpha \pi.$$

The parameters a and b are a priori unknown and are to be determined from experiment. In lowest order we have for the matrix element:

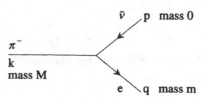

$$\mathcal{M} = \frac{(2\pi)^4 i}{\sqrt{2Vk_0}} \cdot \frac{1}{V} \{\bar{u}(q)\gamma^\alpha(a + b\gamma^5)u(p)\}ik_\alpha.$$

Squaring this matrix element and summing over final spin states (the pion is spinless, so no initial state spin averaging):

$$\sum |\mathcal{M}|^2 = \frac{(2\pi)^8}{8p_0q_0k_0V^3} \ T,$$

$$T = k_\alpha k_\beta \, (-1) \text{Tr} \left[(-i\gamma q + m)\gamma^\alpha(a + b\gamma^5)(-i\gamma p)\gamma^\beta(a^* + b^*\gamma^5)\right].$$

The -1 arises from the complex conjugation and γ^4 movement. In the following we will assume real a and b, thus $a = a^*$, $b = b^*$.

The quantity T will be computed below. First we will do the phase space integration, common to all two-particle decays. The decay width Γ is:

$$\Gamma = \frac{V^3}{(2\pi)^{10}} \int d_3p \, d_3q \, \frac{(2\pi)^8}{8p_0q_0k_0V^3} \, T \, \delta_4(k - p - q)$$

$$= \frac{1}{8(2\pi)^2} \int d_3p \, d_3q \, \frac{1}{p_0q_0k_0} \, T \, \delta_4(k - p - q).$$

Integrating over $\vec{p}$ and the direction of $\vec{q}$ in the k rest system ($k_0 = m_\pi$, $\vec{k} = 0$):

$$\Gamma = \frac{1}{8(2\pi)^2} \cdot 4\pi \cdot \int \frac{q q_0 dq_0}{q_0 p_0 m_\pi} \, T \, \delta(m_\pi - q_0 - p_0)$$

where we used $q \, dq = q_0 \, dq_0$ with $q = |\vec{q}|$ and $q_0 = \sqrt{q^2 + m_e^2}$. In here $\vec{p} = -\vec{q}$. We must find the zero of the argument of the δ-function. Now

$$p_0 = \sqrt{\vec{p}^2 + m_\nu^2} = \sqrt{\vec{q}^2 + m_\nu^2} = \sqrt{q_0^2 - m_e^2 + m_\nu^2}$$

where $m_\nu =$ neutrino mass, and $m_e =$ electron mass. The solution of $p_0 = m_\pi - q_0$ is

$$q_0 = \frac{m_\pi^2 + m_e^2 - m_\nu^2}{2m_\pi}$$

For q we obtain:

$$q = \sqrt{q_0^2 - m_e^2} = \frac{1}{2m_\pi}\sqrt{(m_\pi^2 - m_e^2 - m_\nu^2)^2 - 4m_e^2 m_\nu^2}.$$

The derivative of the argument of the δ-function with respect to q_0 is:

$$\frac{d}{dq_0}(m_\pi - q_0 - p_0) = -1 - \frac{q_0}{p_0} = -\frac{p_0 + q_0}{p_0} = -\frac{m_\pi}{p_0}.$$

where we used the earlier given equation of p_0 in terms of q_0. The result of the q_0 integration is one over the absolute value of this, and the result for Γ is:

$$\Gamma = \frac{1}{8(2\pi)^2} \cdot 4\pi \cdot \frac{q q_0}{q_0 p_0 m_\pi} \, T \, \frac{p_0}{m_\pi}$$

$$= \frac{1}{8\pi} \, q \, \frac{T}{m_\pi^2} \quad \text{(extra factor } \tfrac{1}{2} \text{ if identical particles)}.$$

For future use we noted that a factor $1/2$ is to be included if the two decay products are identical. This is because of Bose statistics. We integrated over all configurations of the outgoing particles, but if they are identical the states obtained by interchange of the two particles are identically the same, and we thus did double counting.

We must now calculate T. Using the fact that γ^5 anticommutes with all other γ we have:

$$(a + b\gamma^5)\,(-i\gamma p)\gamma^\beta(a + b\gamma^5) = (-i\gamma p)\gamma^\beta(a + b\gamma^5)^2$$
$$= (-i\gamma p)\gamma^\beta(a^2 + b^2 + 2ab\gamma^5)$$

The traces may be worked out as usual; note that

$$\text{Tr}(\gamma^\lambda\gamma^\alpha\gamma^\kappa\gamma^\beta\gamma^5) = 4\epsilon_{\lambda\alpha\kappa\beta},$$

and because this is to be multiplied with $k_\alpha k_\beta$ that gives zero due to the antisymmetry of the tensor ϵ. We obtain:

$$T = 4k_\alpha k_\beta\,(a^2 + b^2)\,[q_\alpha p_\beta + q_\beta p_\alpha - \delta_{\alpha\beta}(pq)]$$
$$= 4(a^2 + b^2)\,\Big[2(qk)\,(pk) - k^2(pq)\Big].$$

The various dot-products are:

$$k^2 = -m_\pi^2,\quad q^2 = -m_e^2,\quad p^2 = -m_\nu^2$$
$$(pq) = \tfrac{1}{2}\left((p+q)^2 - p^2 - q^2\right) = \tfrac{1}{2}(k^2 - p^2 - q^2)$$
$$= -\tfrac{1}{2}(m_\pi^2 - m_\nu^2 - m_e^2)$$
$$(qk) = -\tfrac{1}{2}(m_\pi^2 + m_e^2 - m_\nu^2)$$
$$(pk) = -\tfrac{1}{2}(m_\pi^2 + m_\nu^2 - m_e^2)$$

This result, namely that all dot-products can be expressed in the squares of the external momenta, here equal to minus the masses squared of the particles, is common to anything with three external lines. The result for T is:

$$T = 2(a^2 + b^2)\,\{m_\pi^2(m_\nu^2 + m_e^2) - (m_\nu^2 - m_e^2)^2\}.$$

We now specialize to the case at hand, $m_\nu = 0$. Then $q = (m_\pi^2 - m_e^2)/2m_\pi$, and using the earlier given expression for Γ the result is:

$$\Gamma = \frac{a^2 + b^2}{8\pi}\,\frac{(m_\pi^2 - m_e^2)^2}{m_\pi^3}\,m_e^2\,.$$

The same calculation could be done for the decay $\pi \to \mu\nu$, the

only difference being that the electron mass m_e is to be replaced
by the muon mass, to be denoted by m_μ. For the ratio of the
decay rates $\Gamma(\pi \to e\nu)/\Gamma(\pi \to \mu\nu)$ we then find:

$$\frac{\Gamma(\pi \to e\nu)}{\Gamma(\pi \to \mu\nu)} = \frac{(m_\pi^2 - m_e^2)^2}{(m_\pi^2 - m_\mu^2)^2} \cdot \frac{m_e^2}{m_\mu^2}$$

Using $m_\pi = 139.6$ MeV, $m_e = 0.511$ MeV and $m_\mu = 105.7$ MeV
this gives:

$$5.492 \cdot \frac{1}{42787} = \frac{1}{7790} = 1.28 \times 10^{-4}.$$

The experimentally established number is $(1.218 \pm 0.014) \times 10^{-4}$
which agrees very well. A similar situation exists for K^+ decays.
Result: theory 2.57×10^{-5}, experiment $(2.44 \pm 0.11) \times 10^{-5}$. Radiative corrections should be taken into account, but they are
not that easy to calculate because π and K mesons are complicated objects. Note that the above ratios are independent of the
coefficients a and b, it really depends only on the fact that the
interaction contains γ^α and k_α (the pion momentum).

Exercise 5.1 Show that

$$k_\alpha \left(\bar{u}(q)\gamma^\alpha(a + b\gamma^5)u(p) \right)$$

is proportional to the electron mass. Hint: write $k = p + q$ and
use the Dirac equation for the spinors.

Going back to the expression giving the decay rate we may
compute the quantity $c_\pi \equiv \sqrt{a^2 + b^2}$ from the known rate of decay into a μ–ν pair. From $\Gamma_\pi = \hbar/\text{lifetime}$ and multiplying with
the fraction for the decay into a μ–ν pair (almost 100 %, thus
fraction very close to one) one finds $c_\pi = 1.496 \times 10^{-9}\,\text{MeV}^{-1}$.
Similarly for the K meson: $c_K = 4.11 \times 10^{-10}\,\text{MeV}^{-1}$. According to Cabibbo the weak coupling is divided up over non-strange
and strange quark decays according to an angle, now called the
Cabibbo angle. The coefficient for strange quark decay is the sine,
that for non-strange quark decay the cosine of that angle. The
K^+ and π^+ mesons are supposed to be different only in that the s
quark in the K meson is replaced by the d quark in the π meson.

Apart from the s quark being much heavier, making the K much heavier, there are supposedly no other differences in the structure of these mesons. The latter is supposedly due to strong interactions, identically the same for s and d quarks. Then the ratio of the coefficients mentioned should reflect itself in the ratio of the quantities c_π and c_K. It follows that $\tan\theta_c$ equals $c_K/c_\pi \approx 0.275$, which gives $\sin\theta_c \approx 0.275$. A more complete analysis, including analysis of baryon leptonic decays (among them neutron decay) yields $\sin\theta_c \approx 0.221 \pm 0.003$. It should be mentioned that all these very different decays are consistent with a parametrization in terms of only one parameter, the Cabibbo angle. This very important discovery was generalized to the Cabibbo–Kobayashi–Maskawa (CKM) matrix when further quarks were discovered.

5.3 Vector Boson Decay

The decays of the neutral and charged vector bosons of weak interactions have been observed at the big accelerators. Here we will consider the decay into a fermion–anti-fermion. The relevant interaction Hamiltonian is:

$$\mathcal{H} = ig\, V_\mu(\bar\psi_x \gamma^\mu(a + b\gamma^5)\psi_y)$$

describing the decay of a vector boson V into fermions x and y. The relevant diagram is shown in the figure. The calculation has much in common with the pion decay calculation. The matrix element is:

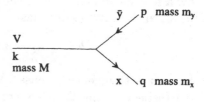

$$\mathcal{M} = \frac{(2\pi)^4 i}{\sqrt{2Vk_0}} \cdot \frac{g}{V} \cdot \{\bar u(q)\gamma^\alpha(a + b\gamma^5)u(p)\}\, ie_\alpha^j(k)$$

where e_α^j is the polarization vector for the V particle. The coefficients a and b are taken to be real. For the matrix element squared, summing over the fermion spins and averaging over the three possible vector boson spin states we find:

$$\frac{1}{3}\sum|\mathcal{M}|^2 = \frac{(2\pi)^8}{8p_0 q_0 k_0 V^3}\, T$$

with

$$T = -\frac{g^2}{3} \left(\delta_{\alpha\beta} + \frac{k_\alpha k_\beta}{M^2} \right).$$
$$\text{Tr} \left[(-i\gamma q + m_x)\gamma^\alpha (a + b\gamma^5)(-i\gamma p - m_y)\gamma^\beta (a + b\gamma^5) \right].$$

The trace may be worked out to give:

$$\text{Tr} = -(a^2 + b^2) \, \text{Tr}(\gamma q \, \gamma^\alpha \gamma p \, \gamma^\beta) - m_x m_y (a^2 - b^2) \text{Tr}(\gamma^\alpha \gamma^\beta)$$
$$= -4(a^2 + b^2)\{q_\alpha p_\beta + p_\alpha q_\beta - (pq)\delta_{\alpha\beta}\} - 4m_x m_y (a^2 - b^2)\delta_{\alpha\beta}$$

Using the dot-product values

$$k^2 = -M^2, \qquad q^2 = -m_x^2, \qquad p^2 = -m_y^2$$
$$(pq) = -\tfrac{1}{2}(M^2 - m_y^2 - m_x^2) \qquad (qk) = -\tfrac{1}{2}(M^2 + m_x^2 - m_y^2)$$
$$(pk) = -\tfrac{1}{2}(M^2 + m_y^2 - m_x^2)$$

the result for T is:

$$T = \frac{2g^2}{3}(a^2 + b^2)\{2M^2 - m_x^2 - m_y^2$$
$$- \frac{(m_x^2 - m_y^2)^2}{M^2}\} + 4g^2 m_x m_y (a^2 - b^2)$$

Now a few special cases. For the decay of the W^- into an electron and an anti-neutrino one has

$$a = b = \frac{1}{2\sqrt{2}}, \qquad m_x = m, \qquad m_y = 0.$$

where m is the electron mass. As the electron mass is very small with respect to the W mass ($m = 0.511$, $M_W = 80220$ MeV) we will set that zero as well. Then

$$T = \frac{2g^2}{3} \cdot \frac{1}{4} 2M^2 = \tfrac{1}{3} g^2 M^2.$$

With these zero masses the momentum $q = M/2$. The result for Γ is:

$$\Gamma = \frac{1}{8\pi} q \frac{T}{M^2} = \frac{1}{8\pi} \frac{M}{2} \frac{g^2}{3} = \frac{g^2}{4\pi} \frac{M}{12}.$$

The numerical result is obtained as follows. The coupling constant g equals the e.m. coupling constant e divided by $\sin \theta_w$, with

θ_w the weak mixing angle, and using

$$\alpha = \frac{e^2}{4\pi} \simeq \frac{1}{137}, \quad \sin^2\theta_w = 0.22$$

the result is

$$\Gamma = \frac{1}{30}\frac{M}{12} = \frac{M}{360}.$$

We included some radiative corrections at these energies by taking a low value for $\sin^2\theta_w$, a more precise value is 0.23. With $M = M_W = 80220$ MeV this gives the result

$$\Gamma_{W \to e\bar{\nu}} = 223 \text{ MeV}.$$

in agreement with the observed data.

For the Z_0 decay into neutrino–anti-neutrino one has

$$a = b = \frac{1}{4\cos\theta_w},$$

while for the decay into electron–positron the values are

$$a = \frac{4\sin^2\theta_w - 1}{4\cos\theta_w}, \quad b = \frac{-1}{4\cos\theta_w}.$$

The Z_0 mass is 91173 MeV, and we leave it to the reader to compare the theoretical decay rates with the observed ones (165 MeV and 83 MeV respectively). At various places radiative corrections are quite important, and a few percent deviation may be expected here.

We conclude with a few remarks on the angular distribution for V decay in the case of a polarized vector boson. Assume that the V is in a definite state of polarization, for example spin up. This in the V rest frame; always go to the rest frame when discussing spin states of a massive particle (if the particle is massless one has two helicity states, spin along and spin opposite to the direction of motion, taken to be along the z-axis). The polarization vector corresponding to this spin up state is $(1, i, 0, 0)/\sqrt{2}$.

Exercise 5.2 Check that this polarization vector is eigen vector with eigenvalue $\lambda = exp(i\phi)$ for rotations over an angle ϕ around the z-axis, i.e. $L_{\mu\nu}e_\nu = \lambda e_\nu$ where L is such a rotation.

The other polarization states, spin down and no spin component along the z-axis are $(1, -i, 0, 0)/\sqrt{2}$ and $(0, 0, 1, 0)$ respectively. They have eigenvalues $exp(-i\phi)$ and $1 = exp(0i\phi)$ for rotations around the z-axis. The coefficient of $i\phi$ in the exponent is the value of the spin component along the z-axis.

In doing the calculation one must now not use the equation for the sum over spins of the e_α^j, but simply keep that vector. The trace calculation must be redone, because the γ^5 part is now important. Taking $m_x = m_y = 0$ the result is:

$$\text{Tr} = -4(a^2 + b^2)\{q_\alpha p_\beta + p_\alpha q_\beta - \delta_{\alpha\beta}(pq)\} - 8ab\,\epsilon_{\mu\nu\alpha\beta}p_\mu q_\nu.$$

Multiplying this with the polarization vector mentioned (and its complex conjugate) one obtains:

$$8abMq_3 - (a^2 + b^2)(M^2 + 4q_3^2)$$

where we used $\vec{p} = -\vec{q}$ and $(pq) = -M^2/2$. For the special case $a = b$ (as in W decay) the result is:

$$-2a^2(M - 2q_3)^2.$$

Writing $q_3 = q\cos\theta$ and $q = M/2$ one obtains the angular distribution:

$$M^2(1 - \cos\theta)^2.$$

This shows a maximum for emission of the electron along the negative third axis, and zero for emission in the opposite direction. In the V-A theory (which corresponds to $a = b$), the spin of a particle (the electron here) is directed opposite to its direction of motion. Therefore a W with spin up emits the electron in the downward direction.

5.4 Muon Decay. Fiertz Transformation

The muon is the first discovered meson, and has a lifetime of 2.197 μsec., which is quite large as particle lifetimes go. Without that particle it would have been very difficult to see the Cabibbo angle in neutron β decay, because the vector coupling constant in that decay is supposedly equal to $\cos(\theta_C)$ times that in μ decay. However, the introduction of the angle was not based on the comparison of μ and β decays. It provided strong encouragement though.

The interaction Lagrangian is:

$$\mathcal{L}_i = \frac{G_F}{\sqrt{2}}\left(\bar{\nu}_\mu\gamma^\alpha(1 + \gamma^5)\mu\right)\left(\bar{e}\gamma^\alpha(1 + \gamma^5)\nu_e\right).$$

We used particle symbols to denote the fields, for example $\psi_e = \bar{e}$. The coupling constant G_F is the Fermi coupling constant, and we have followed the convention of not including the factor $\sqrt{2}$ in G_F, although

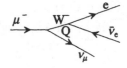

with distaste. In the Standard Model the interaction is mediated by a charged vector boson, with propagator $1/(Q^2+M^2) \approx 1/M^2$. The latter approximation may be safely done since the momentum Q going through the W propagator (of the order of 100 MeV) is very small compared to the vector boson mass M. As the W couples to the (ν_μ, μ) and (ν_e, e) combinations with a factor $g/2\sqrt{2}$ it follows that $G_F/\sqrt{2} = g^2/8M^2$, where g is the coupling constant in the Standard Model. Thus μ decay provides us with a numerical value for g/M.

Denoting the momenta as indicated in the figure, the matrix element is:

$$\mathcal{M} = \frac{(2\pi)^4 i G_F}{V^2\sqrt{2}} \left(\bar{u}_e(p)\gamma^\alpha(1+\gamma^5)u_{\nu_e}(q') \right)$$
$$\left(\bar{u}_{\nu_\mu}(q)\gamma^\alpha(1+\gamma^5)u_\mu(k) \right).$$

The main difference between this calculation and the previous ones is that we now have a three-body phase space to integrate over. In this case there are two particles in the final state with zero mass. We will apply a trick that works very well in case there is at least one particle with zero mass in the final state, and if only the spectrum with respect to one parameter (here the electron energy) is required.

The matrix element squared, summed over final spins and averaged over the muon spin (which amounts to a factor $1/2$), is:

$$\frac{1}{2}\sum|\mathcal{M}|^2 = \frac{(2\pi)^8 G_F^2}{V^4 2^6 k_0 p_0 q_0 q_0'} \left[\gamma^\alpha(1+\gamma^5)\gamma q'(1-\gamma^5)\gamma^\beta \gamma p \right] \cdot$$
$$\left[\gamma^\alpha(1+\gamma^5)\gamma k(1-\gamma^5)\gamma^\beta \gamma q \right].$$

The square brackets imply that the trace must be taken. The mass terms in the spin sums disappear because $(1+\gamma^5)(1-\gamma^5) = 0$.

The first trace:

$$\mathrm{Tr}\left[\gamma^\alpha(1+\gamma^5)\gamma q'(1-\gamma^5)\gamma^\beta\gamma p\right] = 2\,\mathrm{Tr}\left[\gamma^\alpha\gamma q'\gamma^\beta\gamma p(1-\gamma^5)\right]$$
$$= 8(q'_\alpha p_\beta + p_\alpha q'_\beta - \delta_{\alpha\beta}(q'p)$$
$$- \epsilon_{\alpha\lambda\beta\kappa}\,q'_\lambda p_\kappa).$$

The second trace gives a similar expression. Multiplying the two expressions one encounters

$$\epsilon_{\alpha\lambda\beta\kappa}\epsilon_{\alpha\sigma\beta\tau} = 2\delta_{\lambda\sigma}\delta_{\kappa\tau} - 2\delta_{\lambda\tau}\delta_{\kappa\sigma}.$$

Exercise 5.3 Verify this, for example by simply considering the possible values for λ, κ, σ and τ. Remember, there is a summation over α and β, appearing twice in this expression thereby implying summation.

There are only three independent momenta, and no ϵ tensor survives in the result. There is no further difficulty and the result is very simple indeed:

$$\mathcal{M}_s \equiv \frac{1}{2}\sum |\mathcal{M}|^2 = \frac{4(2\pi)^8 G_F^2}{V^4 k_0 p_0 q_0 q'_0}\,(pq)(kq').$$

The expression for the decay rate is:

$$\Gamma = \int d_3 p \int d_3 q \int d_3 q' \left(\frac{V}{(2\pi)^3}\right)^3 \frac{V}{(2\pi)^4}\,\mathcal{M}_s\,\delta_4(k-p-q-qp)$$
$$= \frac{G_F^2}{8\pi^5 k_0} \int \frac{d_3 p}{p_0} \int \frac{d_3 q}{q_0} \int \frac{d_3 q'}{q'_0}\,(pq)(kq')\,\delta_4(k-p-q-q').$$

The integration measures, such as $d^3 p/p_0 = 2d_4 p\,\theta(p_0)\delta(p^2 + m_e^2)$, are Lorentz invariant, and the only non-Lorentz invariant part is the factor $1/k_0$; going from the rest frame to a moving frame, k_0 increases from m_μ to whatever, Γ decreases, and the lifetime, related to $1/\Gamma$, increases as should be. In fact, this is one of the most transparent ways to see the time dilatation factor of the theory of relativity. Now in particular the last part of this expression, containing the integrand and the q and q' integrations is Lorentz invariant. That part will therefore depend only on the various invariants left after the q and q' integrals are done, which are the invariants that can be made from k and p. These invariants are $k^2 = -m_\mu^2$, $p^2 = -m_e^2$ and (kp), from which only the latter is involved in the p integration. Thus all we need to do is to evaluate the q and q' part and obtain it as a function of (kp). The evaluation may be done in any coordinate frame. We will evaluate it in a suitable frame, and then go back to the muon rest frame.

For a given electron momentum p we evaluate the q, q' part in the q, q' rest frame. To be very explicit, let us start in the muon rest frame, and consider a decay with the electron emitted with some momentum p. Take the third axis along the three-momentum $\vec{p}$. Now make a Lorentz transformation along the third axis, boosting in the direction of $\vec{p}$, such that after the Lorentz transformation the muon momentum $\vec{k}$ is equal to the electron momentum. This is possible in general if the momentum transfer to the neutrino system is timelike, i.e., $(q + q')^2 < 0$, which is the case here (except for the limiting case $q = q' = 0$, which is possible for neutrinos).

Exercise 5.4 Verify that $(p + q)^2 \leq 0$ if $p^2 \leq 0$ and $q^2 \leq 0$, and if p_0 and q_0 both positive or both negative.

Note that if the muon three-momentum and the electron three-momentum are equal then the muon energy is larger than the electron energy, which difference is the energy taken by the neutrino pair.

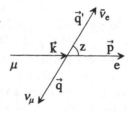

In this frame kinematics is very simple, see figure. The neutrino three-momenta are equal and opposite to each other, $\vec{q}' = -\vec{q}$, thus $|\vec{q}'| = |\vec{q}|$. Since both neutrinos are massless we have in addition $q_0 = |\vec{q}|$ and $q_0' = |\vec{q}'|$. Let E be the energy transferred to the two neutrino system, $E = k_0 - p_0 = q_0 + q_0'$. It follows that $q_0 = q_0' = |\vec{q}'| = |\vec{q}| = E/2$. This very simple result becomes slightly more complicated if one of the neutrino masses is not zero, and if both are non-zero then the result for $|\vec{q}|$ will contain a square root.

On the muon–electron side we have $\vec{k} = \vec{p}$, and the following equation holds:

$$E = k_0 - p_0 = k_0 - \sqrt{|\vec{p}|^2 + m_e^2} = k_0 - \sqrt{k_0^2 - m_\mu^2 + m_e^2}.$$

From this equation k_0 and thus also $p_0 = k_0 - E$ can be solved:

$$k_0 = \frac{E^2 + m_\mu^2 - m_e^2}{2E}.$$

The expression for $|\vec{k}|$ contains a square root, but we do not need it, only $|\vec{k}|^2 = k_0^2 - m_\mu^2$ arises in the expression after we do the integration over $\vec{q}$ directions. Finally we note that $E^2 = -(k-p)^2 = m_\mu^2 + m_e^2 + 2(kp)$, which shows how in this coordinate frame the dependence on (kp) enters.

Everything can be expressed in terms of the variables E and z, where z is the cosine of the angle between $\vec{q}'$ and $\vec{k}$. All dot-products can be written down:

$$(kp) = |\vec{k}|^2 - k_0 p_0 \qquad\qquad (kq') = |\vec{k}||\vec{q}|z - k_0 q_0'$$
$$(kq) = -|\vec{k}||\vec{q}|z - k_0 q_0 \qquad\qquad (pq) = -|\vec{k}||\vec{q}|z - p_0 q_0$$
$$(qq') = -|\vec{q}|^2 - q_0 q_0' \qquad\qquad (pq') = |\vec{k}||\vec{q}|z - p_0 q_0'$$

The integrations over $\vec{q}$ and $\vec{q}'$ are now very easy. The integral over $\vec{q}'$ is trivial and eliminates the three-dimensional part of the δ-function. The result is:

$$\Gamma = \frac{G_f^2}{8\pi^5 k_0} \int \frac{d_3 p}{p_0}$$
$$\int d_3 q \left[k_0^2 + zE|\vec{k}| - z^2|\vec{k}|^2 - Ek_0 \right] \delta(E - q_0 - q_0').$$

Going over to polar coordinates for the neutrino momentum, $d_3 q = |\vec{q}|^2 \, d|\vec{q}| \, dz \, d\phi$, we may do the entirely trivial integrations over ϕ (giving 2π) and z (limits -1 and 1). Note that there is no z dependence in the remaining δ-function for energy conservation, the only z dependence is in the invariant part, and there only terms z and z^2. The integration over $|\vec{q}|$ takes care of the remaining δ-function, $\delta(E - q_0 - q_0')$, with $q_0 = |\vec{q}|$ and $q_0' = |\vec{q}|$. Doing this integration gives a factor $1/2$. That factor is one over the absolute value of the derivative of the argument of the δ-function with respect to $|\vec{q}|$. This δ-function integration factor becomes q_0/E in case the mass of the particle associated with q is non-zero.

All in all, the integrations over $\vec{q}$ and $\vec{q}'$ amount to:
– a factor $1/2$;
– a factor 2π from the $\vec{q}$ azimuth integration;

– replacing z by 0, z^2 by $2/3$;

– terms without z get a factor 2.

Writing $|\vec{k}|^2 = k_0^2 - m_\mu^2$ and substituting for k_0 the expression $(E^2 + m_\mu^2 - m_e^2)/2E$ found above, we have an expression in terms of E, in fact, as it happens, E^2 only. Writing E^2 as function of (kp), we have the desired Lorentz invariant expression in terms of the single invariant (kp).

Let us make this clear. The variable $E^2 = m_\mu^2 + m_e^2 + 2(kp)$ is manifestly Lorentz invariant. In the neutrino rest system it is the energy transferred to the neutrinos. In the muon rest system it is a function of the electron energy obtained by writing $(kp) = -m_\mu p_0$.

At this point we go to the muon rest system, which amounts to substituting $k_0 = m_\mu$ and $(kp) = -m_\mu p_0$ and do the remaining integration over the electron momentum $\vec{p}$. There is no angular dependence, so integrating over angles gives a factor 4π. After that an integral over the magnitude of the electron momentum remains. It is at this point very easy to establish the limits of that integration. In the neutrino pair rest system the neutrino momenta must be positive or zero, which means $E \geq 0$. Minimally then $E = 0$, no energy transfer to the neutrinos in the neutrino rest system. In the muon rest system the condition $E = 0$ solves to $p_0 = (m_\mu^2 + m_e^2)/2m_\mu$. That is the maximum value of the electron energy, which, using $|\vec{p}|^2 = p_0^2 - m_e^2$, gives $(m_\mu^2 - m_e^2)/2m_\mu$ as maximum for the variable $|\vec{p}|$. It is interesting to note that at this endpoint, in the neutrino rest system (!) the muon and electron energies are infinite. It is a limiting case.

We note that up to this point the same treatment can be applied to neutron decay, where there is one particle of zero mass in the final state.

The result is now:

$$\Gamma = \int dp_\ell \, \frac{G_F^2 p_\ell^2}{4\pi^3} \left(m_e^2 - \frac{2m_e^2 m_\mu}{p_0} + m_\mu^2 - \frac{4p_\ell^2 m_\mu}{3p_0} \right),$$

where we have introduced the notation p_ℓ for the absolute value of the electron momentum $|\vec{p}|$. We did something sneaky, namely substituting p_0 as p_0^2/p_0 which is equal to $p_\ell^2/p_0 + m_e^2/p_0$. This is to avoid positive powers of p_0. At this point the only variable left is the electron energy. The above equation gives this spectrum for our choice of the interaction Lagrangian, the V-A interaction. Historically, Michel computed the electron spectrum for a variety of interaction Lagrangians, and parametrized the spectra obtained by means of a few parameters. Having summed over spins, in the limit of zero electron mass, only one parameter remains, ρ_m, called the Michel parameter. For $m_e = 0$ the spectrum found by Michel as function of the electron energy in units of half the muon mass is:

$$\Gamma = \frac{G_F^2 m_\mu^5}{16\pi^3} \int_0^1 dx\, x^2 \left[1 - x + \tfrac{2}{3}\rho_m(\tfrac{4}{3}x - 1)\right] ,$$

with $x = 2p_\ell/m_\mu$. Setting the parameter ρ_m (which has nothing to do with the ρ parameter used in connection with the vector boson mass ratio) equal to 3/4 gives the spectrum that we obtained. The experimental verification of this value for the Michel parameter is then a confirmation of the validity of the V-A theory, as in the Standard Model. Needless to say that experiment has indeed found the value 3/4 for the Michel parameter, with considerable accuracy (0.3%) and taking into account radiative corrections.

We now do the final integration, not neglecting m_e. The essential trick is to introduce the variable $x = p_\ell + p_0$, with $p_0 = \sqrt{p_\ell^2 + m_e^2}$. One has:

$$p_\ell = \frac{x^2 - m_e^2}{2x} ; \qquad \frac{dx}{dp_\ell} = 1 + \frac{p_\ell}{p_0}, \text{ or } \frac{dp_\ell}{p_0} = \frac{dx}{x} .$$

We established already the integration range of p_ℓ, but we will do that again to provide more insight. The minimum is simply zero, the electron remains at rest relative to the muon. The maximum value obtains if both neutrinos are aligned and move opposite to the electron. The sum of the momenta of the neutrinos is then equal to the electron momentum, and because the neutrino's are massless and aligned this is also the total neutrino energy. Since electron plus neutrino energy must be equal to the muon mass we have the equation $p_0 + p_\ell = m_\mu$, which solves to:

$$0 \leq p_\ell \leq \frac{m_\mu^2 - m_e^2}{2m_\mu} .$$

The minimum and maximum values of the variable x are correspondingly m_e and m_μ.

When doing the integration it is actually simplest to use the variable x only for terms involving $1/p_0$. All other terms contain only powers of p_ℓ and are easily integrated. For terms containing $1/p_0$ one simply substitutes $1/x$ for $1/p_0$. The result is:

$$\Gamma = \frac{G_F^2}{\pi^3} \left(\tfrac{1}{192} m_\mu^5 - \tfrac{1}{24} m_e^2 m_\mu^3 \right.$$
$$\left. + \tfrac{1}{8} m_e^4 m_\mu \ln \left(\frac{m_\mu}{m_e} \right) + \tfrac{1}{24} \frac{m_e^6}{m_\mu} - \tfrac{1}{192} \frac{m_e^8}{m_\mu^3} \right).$$

The whole is completely dominated by the first term. The second term is down by a factor $8 m_e^2 / m_\mu^2 \approx 2.10^{-4}$. The final result is:

$$\Gamma = \frac{G_F^2 m_\mu^5}{192 \pi^3}.$$

Numbers follow using the observed lifetime τ_μ of 2.197×10^{-6} sec. That gives for the observed Γ the value $\hbar/\tau = 2.996 \times 10^{-16}$ MeV, which gives $G_F = 1.1638 \times 10^{-11}$ MeV^{-2}.

Electromagnetic radiative corrections can be computed, the result is modified to

$$\Gamma = \frac{G_F^2 m_\mu^5}{192 \pi^3} \left(1 + \delta^{em} \right),$$

with

$$\delta^{em} = -\frac{\alpha}{2\pi} \left(\pi^2 - \tfrac{25}{4} \right) \approx -0.0042.$$

This changes G_F to the value 1.16624×10^{-11}. Not ignoring the first term containing the electron mass makes it 1.16636×10^{-11}. The actual value cited in the particle properties publication 1992 is 1.16639×10^{-11} MeV^{-2}.

It is interesting to reflect on the information contained in verifying the various equations, in particular the value of the Michel parameter ρ_m. The verification gave strong support to the V-A theory, as used here. But from our modern point of view we could pose the following question: is this decay mediated by a vector boson connecting a muon–neutrino pair to an electron–neutrino pair, or perhaps the vector boson is connected on one side to the muon-electron pair and on the other side to the neutrino pair. Is it possible to decide on this on the basis of the equations above? More explicitly, suppose we started from the interaction Lagrangian

$$\mathcal{L}_i = \frac{G_F}{\sqrt{2}} \left(\bar{e}\gamma^\alpha (1 + \gamma^5)\mu \right) \left(\bar{\nu}_\mu \gamma^\alpha (1 + \gamma^5)\nu_e \right) .$$

Would the results be different?

There is of course no problem in actually doing the calculation with this interaction, and pretty soon one discovers that there is no difference. It can also be seen directly for the matrix elements, and since the technique has applications elsewhere we will show how that works. The transformation connecting the two matrix elements is called a Fiertz transformation. It amounts to a reshuffling of spinors and γ-matrices strictly true only in four dimensions, so beware if dimensional regularization is around.

We need one equation from the appendix on traces of γ-matrices, which is the equation that any 4×4 matrix S can be written as a linear combination of γ-matrices:

$$S = a_0 I + a_5 \gamma^5 + a_\kappa \gamma^\kappa + a_\kappa^5 \gamma^\kappa \gamma^5 + a_{\kappa\tau}\sigma^{\kappa\tau} .$$

The indices κ and τ take the values $1\ldots 4$. The matrices $\sigma^{\kappa\tau}$ are the usual anti-symmetric combinations $(\gamma^\kappa\gamma^\tau - \gamma^\tau\gamma^\kappa)/4$, and I is the unit matrix. The 16 coefficients a (the relation $a_{\kappa\tau} = -a_{\tau\kappa}$ holds) are given by:

$$a_0 = \tfrac{1}{4}\text{Tr}[S] \qquad\qquad a_5 = \tfrac{1}{4}\text{Tr}[\gamma^5 S]$$

$$a_\kappa = \tfrac{1}{4}\text{Tr}[\gamma^\kappa S] \qquad\qquad a_\kappa^5 = \tfrac{1}{4}\text{Tr}[\gamma^5\gamma^\kappa S]$$

$$a_{\kappa\tau} = \tfrac{1}{2}\text{Tr}[\sigma^{\tau\kappa} S]$$

We emphasize that the coefficients a are just numbers. Consider now the spinor part of the matrix element for muon decay as it would arise from the new interaction Lagrangian:

$$\left(\bar{u}_e(p)\gamma^\alpha (1 + \gamma^5)u_\mu(k) \right) \left(\bar{u}_{\nu_\mu}(q)\gamma^\alpha (1 + \gamma^5)u_{\nu_e}(q') \right) .$$

From here on we will no more show the momentum dependence of the spinors. Concentrate on the two four-component spinors u_μ and $\bar{u}_{\nu_\mu}$. Together they form a 4×4 matrix, which will be the S in the equation above. Using that equation we obtain:

$$\left(\bar{u}_e\gamma^\alpha (1 + \gamma^5) \left\{ a_0 I + a_5\gamma^5 + a_\kappa\gamma^\kappa \right.\right.$$
$$\left.\left. + a_\kappa^5\gamma^\kappa\gamma^5 + a_{\kappa\tau}\sigma^{\kappa\tau} \right\} \gamma^\alpha (1 + \gamma^5)u_{\nu_e} \right)$$

with

$$a_0 = \tfrac{1}{4}(\bar{u}_{\nu_\mu} u_\mu) \qquad\qquad a_5 = \tfrac{1}{4}(\bar{u}_{\nu_\mu} \gamma^5 u_\mu)$$

$$a_\kappa = \tfrac{1}{4}(\bar{u}_{\nu_\mu} \gamma^\kappa u_\mu) \qquad\quad a_\kappa^5 = \tfrac{1}{4}(\bar{u}_{\nu_\mu} \gamma^5 \gamma^\kappa u_\mu)$$

$$a_{\kappa\tau} = \tfrac{1}{2}(\bar{u}_{\nu_\mu} \sigma^{\tau\kappa} u_\mu)$$

We leave it to the reader to verify by explicitly writing spinor indices that moving $\bar{u}_{\nu_\mu}$ around to the left and putting parenthesis around the expressions gives the same as the trace expressions, for example

$$\text{Tr}[\gamma^5 u\bar{u}] = (\bar{u}\gamma^5 u) = \bar{u}_i \gamma^5_{ij} u_j .$$

The result is the Fiertz transformed expression. Here many terms will be zero, as S has $(1 + \gamma^5)$ on the left and $\gamma^\alpha(1 + \gamma^5) = (1 - \gamma^5)\gamma^\alpha$ on the right. Since $(1 + \gamma^5)X(1 - \gamma^5)$ is zero for $X = I$, $X = \gamma^5$ and $X = \sigma^{\kappa\tau}$ it follows that only the terms with a_κ and a_κ^5 survive. In fact, realizing that $(1 + \gamma^5)(1 - \gamma^5) = 0$ we write

$$a_\kappa \gamma^\kappa + a_\kappa^5 \gamma^\kappa \gamma^5 = \tfrac{1}{2}(a_\kappa + a_\kappa^5)\gamma^\kappa(1 + \gamma^5) + \tfrac{1}{2}(a_\kappa - a_\kappa^5)\gamma^\kappa(1 - \gamma^5)$$

and find that only the combination $\tfrac{1}{2}(a_\kappa - a_\kappa^5)$ survives. That combination is:

$$\tfrac{1}{2}(a_\kappa - a_\kappa^5) = \tfrac{1}{8}(\bar{u}_{\nu_\mu} \gamma^\kappa(1 + \gamma^5)u_\mu).$$

where we exchanged $\gamma^5 \gamma^\kappa = -\gamma^\kappa \gamma^5$ in a_κ^5. The result is:

$$\tfrac{1}{2}(a_\kappa - a_\kappa^5)\left(\bar{u}_e \gamma^\alpha(1 + \gamma^5)\gamma^\kappa(1 - \gamma^5)\gamma^\alpha(1 + \gamma^5)u_{\nu_e}\right) .$$

Using $\gamma^\alpha \gamma^\kappa \gamma^\alpha = -2\gamma^\kappa$ (summation over α implied), and twice $(1 \pm \gamma^5)(1 \pm \gamma^5) = 2(1 \pm \gamma^5)$ the result is, apart from an unobservable sign, the original expression used for the spinor part of the muon decay matrix element:

$$-\left(\bar{u}_{\nu_\mu} \gamma^\kappa(1 + \gamma^5)u_\mu\right)\left(\bar{u}_e \gamma^\kappa(1 + \gamma^5)u_{\nu_e}\right) .$$

5.5 Hyperon Leptonic Decay

Neutron beta decay, $N \to Pe^- \bar{\nu}_e$, is but one example of a whole class of decays, namely hyperon leptonic decays. Others are $\Lambda \to P\ell^-\bar{\nu}$, $\Sigma^+ \to N\ell^+\nu$, $\Sigma^- \to N\ell^-\nu$, $\Sigma^- \to \Lambda\ell^-\bar{\nu}$, $\Xi^- \to \Lambda\ell^-\bar{\nu}$ and $\Xi^- \to \Sigma^0\ell^-\bar{\nu}$. The lepton ℓ may be a muon or an electron. The Cabibbo theory provides relations between these decay modes.

The main task is to compute the integral over the three-body phase space. A full calculation is quite complicated, but can be done along the lines of the muon decay calculation, keeping the electron momentum integration to the last. An electron spectrum can be obtained with relatively few problems. The final integration over the electron momentum is quite nasty, and produces a complicated expression. Unfortunately, the expressions obtained are not convenient for making suitable approximations, and it is for our purposes better to follow other methods.

The decays mentioned above are characterized by the fact that there are two heavy particles involved, whose mass is large with respect to the lepton mass as well as the available phase space. Thus the final lepton momenta are also small compared to these masses. It therefore makes sense to make the corresponding approximations. In hyperon electronic decays one can actually make the approximation of zero electron mass; in neutron decay the available phase space is only 1.3 MeV, and the electron mass is not small compared to this. It should be noted that for hyperon decays phase space is generally of the order of 200 MeV, and the large hyperon mass approximation needs expansion to more than lowest order to obtain a reasonably accurate expression.

In case of hyperon electronic decays, setting the electron mass zero, one can do the calculation of the total decay rate completely along the lines of the muon decay calculation, keeping the proton momentum integration to the last. We will not show that calculation, but quote the result to a sufficient accuracy.

The decay of the neutron is relatively easy to compute, in fact the calculation is on about the same level of complexity as muon decay. The notation for the momenta is shown in the figure. The matrix element is quite similar to that for muon decay:

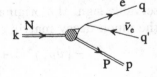

$$\mathcal{M} = \frac{(2\pi)^4 i G_F}{V^2 \sqrt{2}} \left(\bar{u}_e(q)\gamma^\alpha(1+\gamma^5)u_{\nu_e}(q') \right)$$
$$\left(\bar{u}_P(p)\gamma^\alpha(G_v + G_a\gamma^5)u_N(k) \right) .$$

The difference with muon decay is the use of factors G_v and G_a in the proton–neutron piece. The deviation from one for these factors supposedly derives from the complications of the neutron and proton structure in terms of quarks, as well as Cabibbo angle complications. At the bottom of this decay is the transition of the d quark into a u quark and a vector boson (the neutron contains two d and one u quark, the proton two u and one d), and that coupling contains simply $\gamma^\alpha (1 + \gamma^5)$. The complications due to strong interactions may give rise to other terms than those shown above; the most general expression for the proton–neutron part is:

$$\bar{u}_P(p) \left[iG_v\gamma^\alpha + \frac{G_m}{M_n}\sigma^{\alpha\kappa}Q_\kappa + G_sQ_\alpha \right.$$
$$\left. + (iG_a\gamma^\alpha + G_{am}\sigma^{\alpha\kappa}Q_\kappa + G_{ps}Q_\alpha)\,\gamma^5 \right] u_N(k).$$

In here $\sigma^{\alpha\kappa}$ is the familiar combination $(\gamma^\alpha\gamma^\kappa - \gamma^\kappa\gamma^\alpha)/4$, and $Q = k - p$ is the momentum transfer to the leptons. In addition, the various coefficients may be functions of Q^2, and they are real if time-reversal is respected. For neutron decay, where Q is quite small, the terms proportional to Q may be neglected, and also the Q dependence of G_v and G_a may be ignored. For the other hyperon decays the extra terms should in principle be taken into account, but in practice they do not amount to much. It should be mentioned that the coefficient G_m for neutron decay, commonly referred to as weak magnetism, is known. In the Standard Model the CVC hypothesis holds which gives a relation between this coefficient and similar coefficients in photon–neutron and photon–proton coupling. This relation is $G_m = (\mu'_p - \mu'_n)/\sqrt{2}$, where $\mu'_p = 1.793$ and $\mu'_n = -1.913$ are the proton and neutron anomalous magnetic moments respectively.

The momentum dependence of the various coefficients is partly known, partly can be guessed, and is, with one exception, on the scale of 1 GeV and can consequently be ignored as the momentum transfer to the leptons is at best a few hundred MeV. The exception is the coefficient G_{ps} (usually called pseudo-scalar) that supposedly arises mainly due to pion exchange (for example $N \to P\pi^-$ and then $\pi \to e\bar{\nu}$) and can be calculated from the pion–nucleon coupling constant and the pion-decay parameters. This involves the pion propagator, $1/(Q^2 + m_\pi^2 - i\epsilon)$. Because the pion is relatively light, thus inducing a dependence on the scale of 100 MeV, that term could play a role in $\Sigma \rightarrow \Lambda \ell \nu$ decay, with its 70 MeV phase space, but again, it is not worth the trouble. Remember that pion decay into an electron is strongly suppressed; terms proportional to Q_α when multiplied into the lepton part may be seen to be proportional to the mass of the lepton involved just as in the pion decay case. In conclusion we will ignore all but G_v and G_a, and we will also ignore momentum dependence of these factors. We will however quote the contribution of a G_m term where relevant.

It is perhaps useful to mention the fact that the terms $\sigma^{\alpha\kappa}Q_\kappa$ can be rewritten as a linear combination of γ^α and $(k + p)_\alpha$ (with or without γ^5). This may be seen by writing $Q = k - p$, moving γp to the left and γk to the right, and applying the Dirac equation for the products $\bar{u}_P(p)\gamma p$ and $\gamma k\, u_N(k)$. That is useful in calculations, because fewer γ-matrices are involved. Conversely, a term proportional to $(k + p)_\alpha$ may be expressed in γ^α and a σ term. That is the reason that no terms proportional to $(k + p)_\alpha$ are included in the general expression. Here are the precise relations:

$$(\bar{u}_P\sigma^{\alpha\kappa}Q_\kappa u_N) = \tfrac{i}{2}(M_p + M_n)(\bar{u}_P\gamma^\alpha u_N) - \tfrac{1}{2}(k_\alpha + p_\alpha)(\bar{u}_P\, u_N)$$

$$(\bar{u}_P\sigma^{\alpha\kappa}Q_\kappa\gamma^5 u_N) = \tfrac{i}{2}(M_p - M_n)(\bar{u}_P\gamma^\alpha\gamma^5 u_N)$$
$$- \tfrac{1}{2}(k_\alpha + p_\alpha)(\bar{u}_P\gamma^5 u_N)\,.$$

There is a check on these equations: multiplication with Q_α should give zero, because σ is anti-symmetric.

Exercise 5.5 Verify this.

Back to neutron decay, or rather hyperon decay. The first few steps are very similar to those taken in μ decay. The matrix element squared is somewhat more complicated. We obtain in a straightforward manner:

$$\frac{1}{2}\sum |\mathcal{M}|^2 = \frac{(2\pi)^8 G_F^2}{V^4 k_0 p_0 q_0 q_0'}\, T\,,$$

with

$$\begin{aligned}
T = \quad & G_v^2\left[(kq')(pq) + (kq)(pq') + M_n M_p(qq')\right]\\
+ & G_a^2\left[(kq')(pq) + (kq)(pq') - M_n M_p(qq')\right]\\
+ & 2G_v G_a\left[(kq')(pq) - (kq)(pq')\right]\,.
\end{aligned}$$

The trace part T is obviously Lorentz invariant. It depends really only on two independent variables, for which one could take Q^2 and (kq), where $Q = k - p = q + q'$. To see this note first that $q' = k - q - p$ may be eliminated. This leaves three dot-products, namely (kq), (kp), and (pq), in addition to $(kk) = -M_n^2$, $(pp) = -M_p^2$ and $(qq) = -m_e^2$. Now the neutrino mass is zero, thus $0 = (q'q') = (k - p - q, k - p - q)$, and this gives a relation between the three dot-products mentioned. However, we will not go that way.

If one is interested only in the total decay rate then the part with coefficient $G_v G_a$ may be ignored. The reason is that this part is anti-symmetric in the lepton momenta q and q', and when integrating over all lepton configurations this becomes zero. In fact, this part comes from the part of the lepton trace containing $\epsilon_{\alpha\lambda\beta\kappa} q'_\lambda q_\kappa$ (see at muon decay, replacing p by q), and with the constraint that $q + q' = Q = k - p$, this averages out to zero when integrating over lepton configurations. It could however be of relevance when computing the electron spectrum.

We now specialize to neutron decay. The available phase space $\delta_m = M_n - M_p$ (where $M_n = 939.56$ MeV and $M_p = 938.27$ MeV are neutron and proton mass respectively) is only 1.3 MeV, and with the electron mass m_e being only 0.511 MeV it is clear that an approximation in terms of δ_m/M_n and m_e/M_n is quite sufficient. The expression for T may be rewritten as follows:

$$T = (G_v + G_a)^2 \left[(kq')(pq) - (kq)(pq')\right]$$
$$+ 2(G_v^2 + G_a^2)(kq)(pq') + (G_v^2 - G_a^2) M_p M_n (qq') \, .$$

Consider now the decay rate. It is given by:

$$\Gamma = \frac{G_F^2 (2\pi)^8}{2^4 (2\pi)^4 (2\pi)^9 k_0} \int \frac{d_3 q}{q_0} \int \frac{d_3 q'}{q'_0} \int \frac{d_3 p}{p_0} \, T \, \delta_4(k - p - q - q') \, .$$

Let us get some idea concerning magnitudes. In the neutron rest frame the final lepton momenta q and q' are very small, both the three-dimensional parts $\vec{q}$, $\vec{q}'$ and the energies q_0, q'_0. Since the proton three-momentum $\vec{p}$ is equal to minus the sum of the lepton momenta it follows that also $\vec{p}$ is small. The expression for the proton energy, $p_0 = \sqrt{|\vec{p}|^2 + M_p^2}$ differs from the proton mass M_p only by terms of order $|\vec{p}|^2/M_p^2$, i.e., of order δ_m^2/M_n^2, and we may neglect these terms within our approximation. Thus also the proton momentum may be taken as if the proton is at rest. The approximation is the no-recoil approximation.

Consider now the expression T above. To good approximation $(kq) = -M_n q_0$, $(kq') = -M_n q'_0$, $(pq) = -M_p q_0$ and $(pq') = -M_p q'_0$, Furthermore $(qq') = (\vec{q}\vec{q}') - q_0 q'_0$. Inserting this into T we obtain:

$$T = 2(G_v^2 + G_a^2)\, M_n M_p q_0 q'_0$$
$$+ (G_v^2 - G_a^2)\, [M_n M_p |\vec{q}||\vec{q}'|z - M_n M_p q_0 q'_0]\ .$$

In this expression z is the cosine of the angle between the three-vectors $\vec{q}$ and $\vec{q}'$. When integrating over directions, terms proportional to z will average out, and we see that the answer will be proportional to $G_v^2 + 3G_a^2$.

Consider now the equation for the decay rate. First we integrate over proton momenta, eliminating $\delta_3(-\vec{p} - \vec{q} - \vec{q}')$. Everywhere we must now substitute $-\vec{q} - \vec{q}'$ for $\vec{p}$, in particular also in the fourth δ-function. With our approximation we may write M_p for the proton energy. We are left with:

$$\Gamma = \frac{G_F^2}{32\pi^5} \int d_3 q$$

$$\int d_3 q' \left[(G_v^2 + 3G_a^2) + (G_v^2 - G_a^2)\frac{|\vec{q}|}{q_0}z \right] \delta(M_n - M_p - q_0 - |\vec{q}'|).$$

We used that $q'_0 = |\vec{q}'|$ because the neutrino is massless. The integration over $\vec{q}'$ is quite trivial. Put the third axis along the vector $\vec{q}$, and introduce polar coordinates. The integration over the azimuthal angle gives 2π, the integral over z from -1 to 1 gives zero for terms linear in z, and 2 for the z-independent terms. The integral over $|\vec{q}'|$ eliminates the remaining δ-function, forcing $|\vec{q}'| = \delta_m - q_0$, resulting in a factor $(\delta_m - q_0)^2$. Remember, $\delta_m = M_n - M_p$. The integral over directions of $\vec{q}$ is trivial, and gives a factor 4π. The result is now (with $q_0^2 = |\vec{q}|^2 + m_e^2$):

$$\Gamma = \frac{G_F^2(G_v^2 + 3G_a^2)}{2\pi^3} \int dq_\ell$$

$$\left[(m_e^2 + \delta_m^2)q_\ell^2 + q_\ell^4 - 2m_e^2\delta_m\frac{q_\ell^2}{q_0} - 2\delta_m\frac{q_\ell^4}{q_0} \right].$$

where we introduced the notation $q_\ell = |\vec{q}|$ and used $q_0 = q_0^2/q_0$.

This then is the electron spectrum in neutron decay. It agrees well with experiment, for whatever it is worth, as the form of this spectrum appears quite insensitive to the details of the interaction.

The final integration over q_ℓ is very similar to that in muon decay. The terms without $1/q_0$ can be integrated straight away, the integration limits for q_ℓ are 0 and q_m in this approximation. The upper limit for the energy is actually $(M_n^2 - M_p^2 + m_e^2)/2M_n \approx \delta_m + m_e^2/2M_n \approx \delta_m$ from which the limit $q_m = \sqrt{\delta_m^2 - m_e^2}$ can be computed.

Exercise 5.6 Find the upper limit for the electron energy. Hint: for a given electron momentum q go to the proton–neutrino rest frame and determine the limit of the invariant (kq). Then go back to the neutron rest frame. See section on muon decay for more details.

For the terms with $1/q_0$ introduce the variable $x = q_\ell + q_0$; as before, one solves $q_\ell = x/2 - m_e^2/2x$. With $dq_\ell/q_0 = dx/x$ the integration becomes elementary. The limits for the x-integration are m_e and x_m, with $x_m = q_m + \delta_m$. The result is:

$$\Gamma_{N \to Pe\bar{\nu}} = \frac{G_F^2(G_v^2 + 3G_a^2)}{\pi^3}$$

$$\left[\tfrac{1}{60}\delta_m^4 q_m - \tfrac{3}{40}\delta_m^2 m_e^2 q_m - \tfrac{1}{15}m_e^4 q_m + \tfrac{1}{8}\delta_m m_e^4 \ln\left(\frac{q_m + \delta_m}{m_e}\right) \right].$$

To simplify a term containing m_e^8/x_m^4 we used the identity $m_e^2 = 2x_m\delta_m - x_m^2$, readily verified using $q_m^2 = \delta_m^2 - m_e^2$.

There is a check on this equation. Take the equation for muon decay, and set $m_e = M_\mu - \delta_m$. Develop with respect to δ_m, and take the lowest order. The result is the same as the expression above, with $G_v = G_a = 1$ and $m_e = 0$, implying $q_m = \delta_m$.

Numerically this may be used to determine the ratio G_a/G_v, assuming for G_v the value $c_c = 0.975$ (cosine of the Cabibbo angle). From the experimental decay rate of 0.7403×10^{-24} MeV one finds $G_a/G_v = 1.32$. There are however important radiative corrections, notably Coulomb corrections affecting the relatively slow electron as it moves away from the proton. The actual value for $G_a/G_v = 1.257$ according to the 1992 Particle Data table.

For hyperon decay of the type $X \to Ye\nu$ involving an electron or positron one may make the approximation of zero electron mass. From the equation for neutron decay we can immediately find the rate in lowest order, but that is not really good enough. Since in this case $\delta_m/M_x \approx 0.2$ we need one more order to obtain a precision of 5%, and two more orders to get in the 1% range. The calculation can be done similarly to the muon decay calculation, keeping the Y momentum integration to the last. At this point the magnetic term with coefficient G_m can no more be neglected. Here is the result for the total decay rate, up to order δ_m^7:

$$\Gamma_{X \to Yev} = \frac{G_F^2}{\pi^3} \delta_m^5 \left[\frac{G_v^2 + 3G_a^2}{60} \left(1 - \frac{3\delta_m}{2M_x}\right) \right.$$
$$\left. + \frac{G_v^2 + 2G_a^2}{70} \frac{\delta_m^2}{M_x^2} + \left(\frac{G_v G_m}{140} + \frac{G_m^2}{420}\right) \frac{\delta_m^2}{M_x^2} \right].$$

The magnetic term contributes at the 1 % level.

5.6 Pion Decay and PCAC

Pion decay and PCAC make up one of the most interesting subjects of particle physics. It has played a very large role in the discovery of gauge theories, and the anomaly supposedly observed in π^0 decay is to this very day a vexing subject. In this section we will discuss this subject from the modern point of view, i.e., the Standard Model. But make no mistake, there are still many problems around the pion, notably in connection with chiral symmetry.

We will approach the pion as a bound state of quark and anti-quark. Since we truly do not understand this bound state very well we must by necessity be vague at times. The basic idea is this: we note some properties about the appropriate quark-anti-quark combination, and then we assume these properties to hold for the pion.

Let us start with the π^-. It is understood as a bound state of a down quark and an up anti-quark. Consider now the spinors for a spin 1/2 particle at rest. We must build a spinless combination, so let us take a down quark with spin up and an up quark with spin down. From the explicit form of those spinors (namely (1,0,0,0) and (0,0,−1,0)) we see that the Lorentz-invariant combination $\bar{u}d$

is zero. There is one other possibility, namely putting a γ^5 in between. Now γ^5 is of the form

$$\gamma^5 = \begin{bmatrix} 0 & 0 & -1 & 0 \\ 0 & 0 & 0 & -1 \\ -1 & 0 & 0 & 0 \\ 0 & -1 & 0 & 0 \end{bmatrix},$$

and we see that $(\bar{u}\gamma^5 d) = 1$ for the given spin choices. This argument, incidentally, is quite general. The bound state of lowest energy of a fermion and an anti-fermion of spin 1/2 is of the form $(\bar{u}\gamma^5 v)$ or else $(\bar{u}\gamma^\mu v)$. The second possibility gives a state behaving as a vector. For quarks that is supposedly the ρ meson.

Thus, we formally assume:

$$\pi^- \sim i(\bar{u}\gamma^5 d).$$

We have introduced a factor i to have the correct reality properties. That is like the factor i found in the photon–electron coupling.

How far can we draw this similarity? It is really quite simple: one continues till one runs into a problem. So let us move ahead.

First, consider the behaviour under space reflection. To the parity transformation corresponds a transformation of the spinors, in fact they are multiplied with the matrix γ^4, as described earlier. To provide some more insight consider the combination $(\bar{u}\gamma^\mu v)$, which as has been shown in detail, transforms as a vector under a Lorentz transformation. Now make the replacements $u \rightarrow \gamma^4 u$, i.e., $\bar{u} \rightarrow \bar{u}\gamma^4$, and $v \rightarrow \gamma^4 v$. Since γ^4 anticommutes with γ^1, γ^2 and γ^3 it follows that the first three components of $(\bar{u}\gamma^\mu v)$ obtain a minus sign while the fourth component remains unchanged. This is precisely as required for a space reflection.

Since $\gamma^4 \gamma^5 \gamma^4 = -\gamma^5$ we see that the pion obtains a minus sign under space reflection but remains otherwise the same. This has important consequences. The electromagnetic interactions and the strong interactions, described by quantum electrodynamics and quantum chromodynamics respectively, are invariant under space reflection. Therefore any process involving only these forces conserves parity, meaning that the amplitude must remain invariant under this operation. Very sloppily, that means for example that there can be no amplitude for a strong or e.m. process involving an odd number of pions (such as the scattering of two pions giving three pions), as that would change sign under parity. There are complicating factors such as angular momentum, but this is just to illustrate the idea. The conservation of parity is a well established fact for processes involving only strong or e.m. interactions. That the pion is an eigenstate of parity with eigenvalue -1 may be considered as proven to high accuracy. So far we are doing fine.

The vector boson W^- couples also to the $\bar{u}d$ combination. More precisely, the interaction describing an incoming down quark and an anti-up quark annihilating into an outgoing W^- is:

$$\frac{igc_c}{2\sqrt{2}} W_\mu^+ (\bar{u}\gamma^\mu(1+\gamma^5)d).$$

In here $c_c \approx 0.975$ is the cosine of the Cabibbo angle. The decay of the pion into a muon–neutrino pair supposedly proceeds somehow through this interaction. The question is how the pion couples to the W.

Consider the figure. The incoming double line is the pion, the blob is whatever combination of diagrams enter into this process. We know little about it. But one thing is sure: as 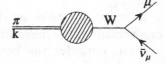 the whole must be Lorentz invariant the blob must behave as a vector. It is a function of the incoming momentum k only, thus it is proportional to the four-vector k_α.

We can also identify some factors coming from the coupling of the W to the quark and lepton combinations, namely $g/2\sqrt{2}$ for each vertex. Also, as here the momentum flowing through the W propagator is of the order of the pion mass, i.e., about 140 MeV, we may ignore that compared to the W mass and the propagator becomes simply $1/M^2$. The diagram gives thus a contribution of the form

$$\frac{(2\pi)^4 i}{V\sqrt{2Vk_0}} \frac{g^2 c_c F(-k^2)k_\alpha}{8M^2} \left(\delta_{\alpha\beta} + \frac{k_\alpha k_\beta}{M^2}\right) (\bar{\mu}\gamma^\beta(1+\gamma^5)\nu_\mu).$$

The combination $F(-k^2)k_\alpha$ supposedly corresponds to everything on the left of the W propagator. The reason why we put a minus sign in front of the argument of F becomes obvious shortly. The quantity c_c is the cosine of the Cabibbo angle.

Here is the argument as to why a function $f_\mu(k)$ that behaves as a vector and depends only on one vector k must be proportional to that vector. First go to the rest frame, where k is of the form $(0, 0, 0, im)$. Now this k is invariant under space rotations, simply because all its space components are zero. Therefore also the function f_μ, depending only on k, must be invariant under space rotations. Therefore its space components must be zero as well. It has thus only the fourth component non-zero, just like k, and it is thus proportional to k. This remains true also after performing any Lorentz transformation because both k and f_μ behave as four-vectors, i.e., transform in the identical way.

This derivation assumes that k is such that one can go to the k rest frame. We leave it to the reader to check the other possibilities.

Exercise 5.7 Show that any function f that is Lorentz invariant and depends only on the momentum k is a function of k^2 only. This is almost a tautology, but at least settles the issue.

We can actually ignore the second term in the W propagator because all components of k are very small compared to the vector boson mass M. Also, the incoming pion is on mass shell, thus we may set $k^2 = -m_\pi^2$. Since the pion decay rate was in fact computed before, we know the value of the function $F(m_\pi^2)$ from comparison with experiment. Referring to the earlier calculation, replacing a and b by $g^2 c_c F(m_\pi^2)/8M^2 = G_F c_c F(m_\pi^2)/\sqrt{2}$ in the matrix element we have:

$$\Gamma_{\pi \to \mu\bar{\nu}} = \frac{G_F^2 c_c^2}{8\pi} \frac{(m_\pi^2 - m_\mu^2)^2}{m_\pi^3} m_\mu^2 \, F(m_\pi^2)^2 \,.$$

The decay makes for practically all of the pion decay. The pion lifetime is 2.6030×10^{-8} sec., which gives a decay rate of $\hbar/\text{lifetime} = 2.52866 \times 10^{-14}$ MeV. With $c_c \approx 0.975$ and $G_F = 1.16639 \times 10^{-11}$ MeV^{-2} one finds $F(m_\pi^2) = 131.52$ MeV, which we will write as $\sqrt{2}f_\pi = 0.94 m_\pi$, charged pion mass $(=139.568$ MeV$)$ intended. One has $f_\pi = 93$ MeV.

Computing the pion decay coupling constant is at this point nothing more than an exercise. But it is possible to go a step further, because there is a trick relating the vector boson coupling to quarks to the pion. The coupling of vector bosons to quarks is as sketched above, i.e., to a combination of the form $(\bar{u}\gamma^\alpha(1+\gamma^5)d)$. The idea is now to study the divergence of this, which is this combination multiplied by ik_α. Consider the diagram part shown in the figure. Leaving out most factors for the moment the corresponding expression will be called a current and denoted by j_α:

$$j_\alpha \equiv i\left(\bar{u}(p)\gamma^\alpha(1+\gamma^5)d(q)\right).$$

The spinors satisfy the Dirac equation for particle and antiparticle:

$$(i\gamma q + m_d)d(q) = 0, \qquad \bar{u}(p)(i\gamma p - m_u) = 0.$$

At this point, as we will apply this work in different circumstances, we will take all momenta directed inwards. Thus $k_\alpha = -q_\alpha - p_\alpha$. Some trivial algebra gives:

$$
\begin{aligned}
ik_\alpha j_\alpha &= -i(q_\alpha + p_\alpha)j_\alpha \\
&= \left(\bar{u}(p)\gamma p(1+\gamma^5)d(q)\right) + \left(\bar{u}(p)\gamma q(1+\gamma^5)d(q)\right) \\
&= -im_u\left(\bar{u}(p)(1+\gamma^5)d(q)\right) + im_d\left(\bar{u}(p)(1-\gamma^5)d(q)\right) \\
&= i(m_d - m_u)\left(\bar{u}(p)d(q)\right) - i(m_u + m_d)\left(\bar{u}(p)\gamma^5 d(q)\right).
\end{aligned}
$$

The first term behaves under reflections as a scalar (invariant under reflections). We concentrate on the second term, behaving as a pseudo-scalar (getting a minus sign under reflection). That part is called the axial part, in other words we concentrate on the divergence of the axial part of the current j_α to which the vector boson couples. This part is what we have decided to call a pion. In line with our strategy we therefore decide that the divergence of the axial part of the current is a constant times the pion field. As we know for pion decay what this axial current is we can even determine the proportionality factor, up to a phase. In the case of pion decay the current is Fk_α, and it absorbs one pion. Its divergence is $ik^2F = -im_\pi^2 F$, which is $-i\sqrt{2}f_\pi m_\pi^2$, and it will still absorb one pion. We therefore find as field equation, essentially the Fourier transform of the above:

$$\partial_\alpha j_\alpha^A(x) = \sqrt{2}f_\pi m_\pi^2 \pi(x).$$

This equation formally expresses the relation between the axial current and the pion field. It is called the PCAC hypothesis (Partially Conserved Axial Current). The quantity m_π is the charged pion mass. In the following we will drop the superscript A.

It may be noted that the quantities in the PCAC equation depend on what one calls the axial current. We took it as $\bar{u}\gamma^\mu\gamma^5 d$, but in the literature one usually takes this divided by $\sqrt{2}$. These are the currents that are used in the algebra of currents of Gell-Mann. The equation often written in the literature is then:

$$\partial_\alpha j^l_\alpha(x) = f_\pi m^2_\pi \pi(x).$$

The quantity $f_\pi = 93$ MeV occurs also in σ-model type Lagrangians describing pion–pion interactions and is then the vacuum expectation value as occurring in that model.

Let us discuss for a moment the "derivation" above. Crucial is that k_α multiplied into the axial part of the current became that combination that we identified as the pion field (apart from a factor). For this we had to use the Dirac equation for the spinors. But the quarks are not really represented by spinors. We have a situation as depicted in the diagram in the figure. Instead of spinors we have quark propagators attached to the vector-boson vertex. Can we still carry the derivation through?

That is the crux of the problem. For the moment we just assume that we can. Within the Standard Model, as a property of a gauge theory, one can find an identity between diagrams, a Ward identity, that precisely states the desired property: you can apply the Dirac equation as if the two quark propagators were spinors. That is not an easy thing to derive, and it can not be done here. Historically one simply assumed this to be correct, and when physics results supported the assumptions, it could be (and was) interpreted as evidence for the existence of Ward identities. That then suggested in turn an underlying gauge theory.

Let us now try to apply our new found knowledge to another situation involving quarks. The prime example is β decay. It proceeds as shown in the figure. It has all the same elements as seen in pion decay, namely the W and the associated lepton pair, and then a coupling to some animal in which quarks are going around. Here that animal is the neutron–proton system as it couples to the W, and that has been well studied experimentally.

Neutron decay as depicted is well described by the interaction Lagrangian

$$\frac{ig}{2\sqrt{2}}\left(\bar{P}\gamma^{\alpha}(G_v + G_a\gamma^5)N\right)W_{\alpha}^{+} \equiv \frac{g}{2\sqrt{2}}j_{\alpha}W_{\alpha}^{+}\,.$$

The constants G_v and G_a are called the vector and axial vector coupling constants respectively. Their values are roughly 1 and 1.25 respectively, and we ignore the 3 % effects of the Cabibbo rotation $(c_c = \cos(\theta_C) \approx 0.975)$.

Now take the divergence of the current coupled to the W, and concentrate on the axial part. See the above figure for the notation for the momenta. Precisely as for the quark case above we find:

$$iQ_{\alpha}j_{\alpha} = i(m_p - m_n)G_v\left(\bar{u}_p(p)u_n(k)\right)$$
$$+i(m_p + m_n)G_a\left(\bar{u}_p(p)\gamma^5 u_n(k)\right)\,.$$

The first term may be ignored, as the neutron and proton mass are very nearly equal. We are left with the axial piece.

Consider the coupling of a pion to proton and neutron. That is well described by the interaction

$$G_{\pi N}\left(\bar{P}\gamma^5 N\right)\pi\,.$$

The pion–nucleon coupling constant is well known from experiment: $G_{\pi N} = 19.16$. The divergence of the axial current is assumed to behave as a pion with a factor $\sqrt{2}f_{\pi}m_{\pi}^2$, and that couples then to the nucleons as shown in the diagram. The momentum flowing through the pion line is of the order of the nucleon mass difference, ≈ 1.29 MeV, and it may be ignored compared to the pion mass. We may therefore take simply $1/m_{\pi}^2$ for the pion propagator. The identity that we now have is:

$$(m_p + m_n)G_a = \frac{G_{\pi N}^{'}}{m_{\pi}^2}\sqrt{2}f_{\pi}m_{\pi}^2\,.$$

This may be used to compute G_a numerically, using $G_{\pi N}$ as input. The result is $G_a = 1.34$, which is not that far from the experimental result. It would have appeared a lot more spectacular if we had computed $G_{\pi N}$ from G_a, and it must be realized how big this pion–nucleon coupling constant must be to generate the correct G_a. Given the nature of the derivation it must really be considered a succes.

The relation derived here is called the Goldberger–Treiman relation, after the discoverers. It is a quite impressive achievement if this is seen against the state of affairs at that time (1958).

A more refined derivation concerning the axial vector coupling constant is due to Adler and Weisberger. It involves consideration of a process with two axial currents, something like vector-boson nucleon scattering. Then there is a modification of the simple PCAC relation proposed above, which in terms of Ward identities of a gauge theory can actually be understood. This can be discussed only in the context of the full Standard Model, with a detailed knowledge of Ward identities. The Adler–Weisberger relation is a relation between G_a and the pion–nucleon scattering length. It is well satisfied.

5.7 Neutral Pion Decay and PCAC

The decay of the neutral pion into two photons has played a significant role in the development of the Standard Model, and as it supposedly proceeds through the anomaly it is still of considerable interest.

We begin with the phenomenology of the decay. The vertex factor for the diagram shown is easily established. Since under space reflections the pion changes sign, while the pair of photons, no matter what they do, can make no sign change we must introduce an additional quantity that changes sign under reflections. Then the matrix element will remain invariant, as it should for this process where only strong and e.m. interactions are involved. For this one has the ϵ tensor. For this decay there happen to be four independent vectors, namely the momenta and polarization vectors of the two photons. These vectors are needed to saturate the ϵ indices. We find that the vertex must be of the form:

$$8if_0\epsilon_{\alpha\beta\lambda\kappa}e_\alpha e'_\beta q_\lambda q'_\kappa.$$

The reason for the factor 8 is that this vertex factor corresponds to the interaction Lagrangian:

$$i f_0 \pi \epsilon_{\alpha\beta\lambda\kappa} F_{\alpha\beta} F_{\lambda\kappa}, \quad \text{with} \quad F_{\mu\nu} = \partial_\mu A_\nu - \partial_\nu A_\mu.$$

Exercise 5.8 Verify this.

By computing the decay rate, and comparing with experiment the quantity f_0 may be calculated. Referring to the calculation of the charged pion decay we have:

$$\Gamma_{\pi^0 \to 2\gamma} = \frac{1}{16\pi} q \frac{T}{m_{\pi^0}^2}.$$

An extra factor $1/2$ has been included because the two final particles are identical, and we must correct for double counting in the phase space integration. In the above $q = m_{\pi^0}/2$ and

$$T = -64 f_0^2 \sum \left(\epsilon_{\alpha\beta\lambda\kappa} e_\alpha e'_\beta q_\lambda q'_\kappa \right) \left(\epsilon_{\alpha'\beta'\lambda'\kappa'} e_{\alpha'} e'_{\beta'} q_{\lambda'} q'_{\kappa'} \right).$$

The minus sign arises because the ϵ tensor must be considered imaginary, as it forces one and only one of the indices to be four. That issue has been discussed at length before.

The sum is over photon polarizations, which amounts to the substitution $e_\alpha e_{\alpha'} = \delta_{\alpha\alpha'}$ and similarly for e'. Using $q^2 = q'^2 = 0$ and $(qq') = k^2/2 = -m_{\pi^0}^2/2$ the result is $T = 32 f_0^2 m_{\pi^0}^4$. The decay rate is then:

$$\Gamma_{\pi^0 \to 2\gamma} = \frac{f_0^2 m_{\pi^0}^3}{\pi}.$$

Using the observed lifetime of 8.4×10^{-17} sec. (error about 7%), and taking into account the branching ratio of 98.8% for the decay into two photons we find a decay rate of 7.7×10^{-6} MeV. This gives $f_0 = 3.14 \times 10^{-6}$ MeV^{-1}. This value is to be compared with the value obtained by evaluating the anomaly (to be done later), namely

$$f_0^{\text{anomaly}} = \frac{\alpha}{8\pi f_\pi} = 3.12 \times 10^{-6} \, \text{MeV}^{-1},$$

where $\alpha = e^2/4\pi = 1/137.036$ is the fine-structure constant.

Now that the phenomenology is out of the way we will argue that the decay rate should be zero, evidently in contradiction to the result just obtained. Here is the argument.

Since isospin holds well among pions one may generalize the PCAC equation. Before, the equation was used for the charged pion, involving the charged axial vector current (the axial current coupled to the charged vector boson). We thus assume also the PCAC equation for the neutral axial vector current:

$$\partial_\alpha j_\alpha^0 = \sqrt{2} f_\pi m_\pi^2 \pi^0 .$$

There is the important question if this equation, obtained from the equation for the charged case, remains true including e.m. interactions. Here we may argue as follows. Gauge invariant e.m. interactions are obtained by substituting $\partial_\mu - iqA_\mu$ for ∂_μ, where q is the charge of the object on which the derivative operates. This recipe is quite general, but we will not go into any further details here. According to this recipe, including e.m. interactions affects the charged PCAC relation, but not the neutral one.

An important step forward was taken by Adler, who for a change wrote the PCAC equations in the opposite way:

$$\pi = \frac{1}{\sqrt{2} f_\pi m_\pi^2} \, \partial_\alpha j_\alpha .$$

This equation is now seen as an equation telling us something about the pion field. A derivative implies a momentum in the diagrams. The conclusion that we can draw from the PCAC equation is this: a pion field is like some quantity multiplied by the momentum flowing into that. That quantity, the axial current, is not singular if the associated momentum tends to zero. Therefore we must conclude that any process involving a pion must go to zero if the momentum of the pion tends to zero.

This is Adler's consistency condition. It may be applied to pion–nucleon scattering, but we will consider the application to processes involving a π^0. As argued above, it will then hold even taking into account e.m. interactions.

There is an important but unfortunately quite vague point here. Simply for dimensional reasons expressions will not just have factors k, where k is the pion momentum, but rather k/λ, where λ is some mass scale arising in the process considered. Can we say anything about typical values for this scale? This remains a question of hope and belief. If you think that λ could typically be something like the pion mass then there is no need for you to read further. But the more reasonable value is something like the ρ meson mass, which is 770 MeV, i.e., of the order of 1000 MeV. The point is that in the context of chiral invariance the pion mass is supposedly near zero. But if one forgets about the pion mass then all hadron masses are in the 1000 MeV range. Indeed, experience with energy dependence in processes involving hadrons suggests this kind of energy scale. If the momentum is of the order of magnitude of 100 MeV then we think correspondingly of a suppression by a factor of 0.1, or, since momenta usually occur not single but in dot-products, more like a factor 0.01. That is for the matrix element, and for the observed cross sections or decay rates the suppression factor becomes 10^{-4}.

It must be noted that at various places we tacidly ignore the possible momentum dependence of the factors f_π and f_0. They occur in the actual decay rate, where the momentum squared is $-m_\pi^2$, but also in the equations for the divergence of the axial current where usually the value for zero momentum is needed. It is in line with the argument above to ignore the momentum dependence over this small region, 0–m_π.

The first process to which this argument was applied (by Sutherland) is the decay of the η meson into three pions: $\eta \to \pi^+\pi^-\pi^0$. We expect this decay to be strongly suppressed, and we expect in the study of the process a decreasing rate as function of the π^0 momentum. Nothing of that kind is observed, in fact the opposite is seen. For a long time this has been considered a puzzle, and the issue remains very complicated.

There is also a decay of the η into three neutral pions which should really be suppressed, but which proceeds at a rate roughly 1.5 times that of the charged mode. We will leave it at that.

The very interesting case is π^0 decay. This has to be considered in some detail, because there is already momentum dependence in the interaction, and one might think that is enough to satisfy the PCAC constraint. But that is not so.

Let us first consider how the axial current would look like in this case. Thus imagine a neutral vector boson coupled to two photons. The axial part of interaction would again involve an ϵ tensor. Gauge invariance requires that the amplitude is zero if the polarization vector of any of the photons is replaced by its momentum. Well, here is the simplest thing one can make:

$$j_\mu^a = f_x p_\mu \epsilon_{\alpha\beta\lambda\kappa} e_\alpha e'_\beta q_\lambda q'_\kappa.$$

where one would have to attach a factor e_μ^v if this would indeed describe decay of a vector boson. Here e^v is then the polarization vector associated with the vector boson, and k, q and q' are the vector boson and photon momenta respectively. The vector p_μ is any linear combination of k, q and q', and f_x represents whatever else is there, such as a scale to provide the correct dimensions. Now note that $(kq) = (kq') = 0.5k^2$, i.e., k^2 is the only non-zero quantity. Taking the divergence of the axial current, i.e., multiplying by k_μ, we obtain an extra factor k^2 as compared to the matrix element given in the beginning:

$$k^2 f_x \epsilon_{\alpha\beta\lambda\kappa} e_\alpha e'_\beta q_\lambda q'_\kappa.$$

We therefore conclude that the matrix element for the divergence of the axial current to become two photons is strongly suppressed. It would therefore appear that if PCAC is any good then also π^0 decay is strongly suppressed. To settle this quantitatively let us assume that PCAC was actually not as written, but that for the π^0 case there would be an extra term:

$$\partial_\alpha j_\alpha^0 = \sqrt{2} f_\pi m_\pi^2 \pi^0 + R.$$

where R is an unknown quantity involving whatever fields may be around. Consider this equation in terms of diagrams, see figure.

If we take the divergence of the axial current (the left hand side of the equation) to be zero then we get a relation between the quantity R and the actually observed pion decay. Inserting for the π^0 coupling to the two photons the phenomenological expression derived before, we obtain for the first term on the right hand side of the equation the expression

$$\frac{\sqrt{2} f_\pi m_\pi^2}{k^2 + m_{\pi^0}^2 - i\epsilon} 8i f_0 \epsilon_{\alpha\beta\lambda\kappa} e_\alpha e'_\beta q_\lambda q'_\kappa.$$

The second term depends on R. At the very least R should contain two photon fields, to emit the two outgoing photons. Since the R term must be something like the opposite of the first term the simplest assumption for R is:

$$R = 8i c_a \epsilon_{\alpha\beta\lambda\kappa} e_\alpha e'_\beta q_\lambda q'_\kappa,$$

with c_a an unknown constant. Such an expression results if in the PCAC equation one writes

$$\partial_\alpha j^0_\alpha = \sqrt{2} f_\pi m_\pi^2 \pi^0 + i c_a \epsilon_{\alpha\beta\lambda\kappa} F_{\alpha\beta} F_{\lambda\kappa}.$$

It should be noted that using m_π or m_{π^0} is a matter of taste. The difference, about 3%, is an isospin breaking effect, and that we have not under control.

Let us now consider the resulting equation in the limit of zero momentum k of the pion. Then we may ignore the momentum in the pion propagator. The equation becomes an equation for the constant c_a, and one may solve $c_a = -\sqrt{2} f_0 f_\pi$. When computing the anomaly we will obtain $c_a = \sqrt{2}\,\alpha/8\pi$. That, as already noted above, agrees very well with experiment.

Unfortunately, the η decay into three pions is not cured by the R term.

6

Renormalization

6.1 Introduction

The treatment in this chapter is sketchy, and not meant to be complete. The main purpose is to show the importance of the behaviour of the vector boson propagator with respect to renormalizability of the theory. In the foregoing the important equation

$$\sum_i e^i_\mu(k)\bar{e}^i_\nu(k) = \delta_{\mu\nu} + \frac{k_\mu k_\nu}{\kappa^2}$$

for the polarization vectors of the (massive) electromagnetic field was given. The time-ordered product of the e.m. field contains these polarization vectors, and this leads to a propagator for the e.m. field that differs from the scalar case (σ or π field) by just this factor:

$$\frac{\delta_{\mu\nu} + k_\mu k_\nu/\kappa^2}{k^2 + \kappa^2 - i\epsilon}.$$

For large values of k this behaves as a constant, rather than as $1/k^2$ for the σ or π case, and with respect to renormalizability this is of crucial importance. In quantum electrodynamics the $k_\mu k_\nu/\kappa^2$ part may be dropped altogether. In gauge theories in general, such as the Standard Model, one can show that this part may be dropped in exchange for certain other better behaving terms, and that demonstration is really at the nucleus of the proof of renormalizability of such theories.

6.2 Loop Integrals

Let us go back to the very simple π–σ model, with the interaction Hamiltonian

$$\mathcal{H} = g\pi^2\sigma.$$

137

We considered earlier the cross section for π–π scattering; one of the contributing diagrams involves σ exchange
in the t-channel (meaning that the
momentum q flowing through the
σ-line is related to the momentum
transfer, i.e., $t = -Q^2 = -(k - k')^2$).

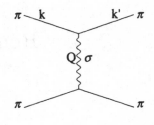

In higher order other diagrams occur. They will generally involve an integral over some loop momentum. An example is shown
in the figure, there is a σ self-energy insertion.

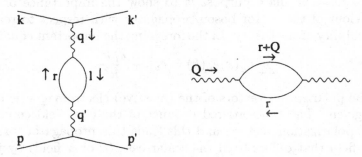

We will leave out all factors common to this diagram and the
lowest order diagram above, i.e. the external line factors and
an overall factor $g^2(2\pi)^4 i$. The expression corresponding to this
diagram is then:

$$\tfrac{1}{2} \int d_4q \int d_4q' \int d_4r \int d_4\ell$$

$$\frac{4g^2/(2\pi)^4 i}{(q^2 + m^2 - i\epsilon)\,(q'^2 + m^2 - i\epsilon)\,(\ell^2 + M^2 - i\epsilon)\,(r^2 + M^2 - i\epsilon)}$$

$$\delta_4\,(k - k' - q)\,\delta_4\,(q + r - \ell)\,\delta_4\,(\ell - r - q')\,\delta_4\,(q' + p - p')\,.$$

The factor $\tfrac{1}{2}$ is a combinatorial factor discussed before for this
diagram. The derivation can be found in the appendix on combinatorial factors. Doing the q, q' and ℓ integral, and introducing
the notation $Q = k - k' = p' - p$ the expression reduces to

$$\frac{2g^2}{(2\pi)^4 i}\, I(Q)\, \frac{1}{(Q^2 + m^2 - i\epsilon)^2}\delta_4(k - k' + p - p')$$

with the r-dependent part

$$I(Q) = \int d_4r \frac{1}{((r+Q)^2 + M^2 - i\epsilon)(r^2 + M^2 - i\epsilon)}.$$

This corresponds to the self-energy insertion shown in the figure (apart from the combinatorial factor $\frac{1}{2}$). Let us now concentrate on evaluating this function $I(Q)$. As it happens the r-integral is infinite, because for large r it behaves as

$$\int d_4r \frac{1}{r^4}$$

which diverges logarithmically. Let us consider this in detail.

First one uses the Feynman trick for combining denominators

$$\frac{1}{ab} = \int_0^1 dx \frac{1}{(ax + b(1-x))^2}.$$

This equation can be verified by direct calculation:

$$= \int_0^1 dx \frac{1}{(x(a-b)+b)^2} = \frac{-1}{a-b}\left[\frac{1}{x(a-b)+b}\right]_0^1$$

$$= -\frac{1}{a-b}\left[\frac{1}{a} - \frac{1}{b}\right] = \frac{1}{ab}.$$

Using this we have

$$I(Q) = \int_0^1 dx \int d_4r \frac{1}{[x(r+Q)^2 + (1-x)r^2 + M^2 - i\epsilon]^2}$$

$$= \int_0^1 dx \int d_4r \frac{1}{[r^2 + 2x \cdot (rQ) + xQ^2 + M^2 - i\epsilon]^2}.$$

The next step is to replace the integration variable r by the variable q, with $q = r + xQ$. This is chosen such that the term in the denominator linear in the integration variable disappears:

$$I(Q) = \int_0^1 dx \int d_4q \frac{1}{[q^2 + A - i\epsilon]^2}$$

where we used the abbreviation $A = x(1-x)Q^2 + M^2$.

To proceed with this integral we first perform the Wick rotation, which transforms the integration from Minkowski to Euclidean space. Remember that

$$\int d_4q = \int_{-\infty}^{\infty} dq_0 \int d_3q$$

and $q^2 = \vec{q}^2 - q_0^2$. Consider now the denominator in the integral

$$q^2 + A - i\epsilon = \vec{q}^2 + A - i\epsilon - q_0^2.$$

Let us for definiteness assume that $\vec{q}^2 + A$ is positive. The denomina- tor will be zero for

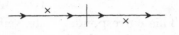

$$q_0 = \pm\sqrt{\vec{q}^2 + A - i\epsilon} = \pm\left\{\sqrt{\vec{q}^2 + A} - i\epsilon\right\}$$

where one really has $i\epsilon/2\sqrt{}$, but that is irrelevant in the limit $\epsilon = 0$. In the complex q_0 plane the integrand has poles at the locations given by this equation and shown in the figure. The integration is along the real axis, from $-\infty$ to $+\infty$.

Consider now the integration contour shown in the figure, drawn at a much smaller scale. The idea is to take the limit where the segments are at infin- ity, i.e., the limit $R \to \infty$ where R is the radius of the circle seg- ments. The total integral over this closed contour is zero be- cause of the fact that there are no poles inside. Furthermore, the denominator drops like q_0^{-4}

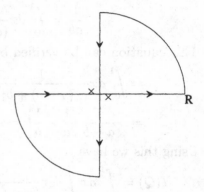

for large q_0, and therefore the integral over the quarter segments will go to zero in the limit where these segments are moved out- wards (the value will be of the order of the length of the segment, $\pi R/2$, times the average value of the integrand, $1/R^4$, over this segment).

If the total integral is zero, and if the integral over the circle segments is zero, then the integral from $-\infty$ to ∞ (along the real axis) must be minus the integral from $+i\infty$ to $-i\infty$ (from the top down along the imaginary axis). Thus, symbolically

$$\int_{-\infty}^{\infty} dq_0 = -\int_{i\infty}^{-i\infty} dq_0 = -\frac{1}{i}\int_{-\infty}^{\infty} dq_4 = i\int_{-\infty}^{\infty} dq_4$$

with $q_4 = iq_0$. Moreover, $\vec{q}^2 - q_0^2 = \vec{q}^2 + q_4^2 = q_1^2 + q_2^2 + q_3^2 + q_4^2$ which is now a Euclidean dot-product. Note that q_4 now takes real values only (from $-\infty$ to $+\infty$).

Thus now:

$$I(Q) = i \int_0^1 dx \int d_4q \frac{1}{[q^2 + A - i\epsilon]^2}, \quad A = x(1-x)Q^2 + M^2$$

with Euclidean q.

Exercise 6.1 Show that the above reasoning also holds if $\vec{q}^2 + A < 0$. Find the location of the poles for that case, and check that they are outside the contour used.

The integral over q can be done easily. First introduce polar coordinates (in four dimensions, see below for details) and integrate over all angles (there is no angular dependence in the integrand). The result is

$$I(Q) = i \int_0^1 dx \int_0^\infty \omega^3 d\omega \frac{2\pi^2}{[\omega^2 + A - i\epsilon]^2}.$$

For the moment we will use the upper limit Λ instead of ∞ for the ω integral; with $z = \omega^2$ and $dz = 2\omega d\omega$ one has

$$I(Q) = i\pi^2 \int_0^1 dx \int_0^{\Lambda^2} dz \frac{z}{[z + A - i\epsilon]^2}$$

$$= i\pi^2 \int_0^1 dx \int_0^{\Lambda^2} dz \left\{ \frac{1}{z + A - i\epsilon} - \frac{A}{(z + A - i\epsilon)^2} \right\}$$

$$= i\pi^2 \int_0^1 dx \left\{ \ln(z + A - i\epsilon) + \frac{A}{(z + A - i\epsilon)} \right\}_0^{\Lambda^2}$$

$$\simeq i\pi^2 \int_0^1 dx \left\{ \ln\left(\Lambda^2 + A - i\epsilon\right) - \ln(A - i\epsilon) - 1 \right\}$$

$$\simeq i\pi^2 \left\{ \ln \Lambda^2 - 1 - \int_0^1 dx \ln(A - i\epsilon) \right\}$$

$$= i\pi^2 \{\ln \Lambda^2 - 1 - \ln(A - i\epsilon)\}$$

where we neglected A compared to Λ^2 and omitted terms behaving as $1/\Lambda^2$. This answer becomes logarithmically infinite as $\Lambda \to \infty$. We say that the integral $I(Q)$ is logarithmically divergent.

We will now compute the somewhat more general integral

$$I(A, \alpha, n) = i \int d_n q \frac{1}{[q^2 + A - i\epsilon]^\alpha}$$

where q is in n-dimensional Euclidean space, $q^2 = q_1^2 + q_2^2 \ldots + q_n^2$. Introduce polar coordinates into this n-dimensional q-space:

$$\int d_n q = \int_0^\infty \omega^{n-1} d\omega \int_0^{2\pi} d\theta_1 \int_0^\pi \sin\theta_2 d\theta_2 \int_0^\pi \sin^2\theta_3 d\theta_3 \ldots$$
$$\int_0^\pi \sin^{n-2}\theta_{n-1} d\theta_{n-1}.$$

For $n = 3$ that is of course a well-known expression, following from $q_1 = \omega \sin\theta \cos\varphi$, $q_2 = \omega \sin\theta \sin\varphi$ and $q_3 = \omega \cos\theta$, with $\varphi = \theta_1$, and $\theta = \theta_2$. In four-dimensional space one has $q_1 = \omega \sin\theta_3 \sin\theta_2 \cos\varphi$, $q_2 = \omega \sin\theta_3 \sin\theta_2 \sin\varphi$, $q_3 = \omega \sin\theta_3 \cos\theta_2$ and $q_4 = \omega \cos\theta_3$.

Exercise 6.2 If you have nothing better to do work out the Jacobian for the case $n = 4$, i.e., the determinant of the matrix J_{ij} with $J_{ij} = dq_i/dx_j$, and $x_1 = \omega, x_2 = \varphi, x_3 = \theta_2$, and $x_4 = \theta_3$.

There is no angular dependence in the integrand, and we can use

$$\int_0^\pi \sin^m\theta d\theta = \sqrt{\pi}\, \frac{\Gamma\left(\frac{m+1}{2}\right)}{\Gamma\left(\frac{m+2}{2}\right)}$$

where $\Gamma(z)$ is the well-known Γ-function, with $z\Gamma(z) = \Gamma(z+1)$. For integer values of z one has $\Gamma(z) = (z-1)!$, in particular $\Gamma(1) = 1$. For $m = 0$ the result must be π, which tells you that $\Gamma(\frac{1}{2}) = \sqrt{\pi}$.

Exercise 6.3 Verify the equation for the case $m = 1$. Establish a recursion relation, by writing $\sin^m\theta = -\sin^{m-1}\theta\, d\cos\theta/d\theta$ and performing partial integration.

With this one obtains:

$$I(A, \alpha, n) = \frac{2i\pi^{n/2}}{\Gamma\left(\frac{n}{2}\right)} \int_0^\infty \omega^{n-1} d\omega \frac{1}{[\omega^2 + A - i\epsilon]^\alpha}.$$

Also this integral is in the books, the result is:

$$I(A, \alpha, n) = \frac{2i\pi^{n/2}}{\Gamma\left(\frac{n}{2}\right)} \cdot \frac{1}{2} \cdot \frac{\Gamma\left(\frac{n}{2}\right)\Gamma\left(\alpha - \frac{n}{2}\right)}{\Gamma(\alpha)A^{\alpha - n/2}}$$

$$= \frac{i\pi^{n/2}}{A^{\alpha - n/2}} \cdot \frac{\Gamma\left(\alpha - \frac{n}{2}\right)}{\Gamma(\alpha)}.$$

This equation holds for any α and any integer value of n such that the original integral exists (i.e., is convergent). This result looks rather different from the previously found result for $n = 4$ and $\alpha = 2$. But for these values we encounter the Γ-function with argument 0, which is infinite. To see clearly what happens take $\alpha = 2$ and n equal to $4 + d$ in the above equation. One has:

$$\Gamma\left(\alpha - \frac{n}{2}\right) \Rightarrow \Gamma\left(-\frac{d}{2}\right).$$

Now $z\Gamma(z) = \Gamma(z + 1)$, i.e., $\Gamma(z) = \Gamma(z + 1)/z$, thus

$$\Gamma\left(-\frac{d}{2}\right) = -\frac{2}{d}\Gamma\left(1 - \frac{d}{2}\right) = -\frac{2}{d}\left(1 + \frac{d}{2}\gamma + \mathcal{O}\left(d^2\right)\right)$$

for small d. In here $\gamma = $ Euler's constant. Further,

$$A^{\frac{d}{2}} = \left(e^{\ln A}\right)^{\frac{d}{2}} = e^{\frac{d}{2}\ln A} = 1 + \frac{d}{2}\ln A + \mathcal{O}\left(d^2\right)$$

for small d. Similarly,

$$\pi^{n/2} = \pi^2 \cdot \pi^{\frac{d}{2}} = \pi^2\left(1 + \frac{d}{2}\ln\pi + \mathcal{O}\left(d^2\right)\right).$$

Altogether we obtain

$$I\left(A, 2, 4 + d\right) = i\pi^2\left(1 + \frac{d}{2}\ln\pi\right)\left(1 + \frac{d}{2}\ln A\right) \cdot$$

$$\left(-\frac{2}{d}\right)\left(1 + \frac{d}{2}\gamma\right) + \mathcal{O}\left(d\right)$$

$$= -\frac{2i\pi^2}{d} - i\pi^2\left(\gamma + \ln\pi\right) - i\pi^2\ln A + \mathcal{O}\left(d\right).$$

This now looks rather similar to the previously obtained result if we identify

$$\ln\Lambda^2 \leftrightarrow 1 - \frac{2}{d} - \gamma - \ln\pi.$$

It appears that both methods give the same, they are just different ways to handle a divergent integral. In both cases the result is infinite: $\ln\Lambda^2$ for $\Lambda \to \infty$ or $2/d$ for $d \to 0$, with $d = n - 4$.

The second method is called the method of dimensional regularization. Formally, at the end of the calculation the dimension is taken as a continuous parameter. The method relies on the continuous function $\Gamma(z)$, originally defined only for integer z through $\Gamma(z) = (z-1)!$.

Using the $\overline{MS}$ subtraction scheme amounts to throwing away the combination $2/d + \gamma + \ln \pi$.

While the calculation of the integrals above may become horribly complicated for more complicated diagrams, it is relatively simple to study the divergent part of these integrals. In first approximation this is done through power counting. That is not an exact method, at least not if going beyond one loop diagrams, but the results and insights obtained are substantially correct.

The power counting method amounts to estimating the degree of divergence simply by considering the leading behaviour for large momentum. Thus $I(Q)$, as argued before, behaves as

$$\int d_4q \, \frac{1}{q^4}$$

which is logarithmically divergent. Integrals behaving like

$$\int d_4q \, \frac{1}{q^3}, \int d_4q \, \frac{1}{q^2} \text{ etc.}$$

are called linearly, quadratically etc. divergent. For example:

$$I_2 = \int d_4q \frac{1}{[q^2 + M^2 - i\epsilon]}$$

may be worked out by methods as above to give:

$$I_2 = i\pi^2 \left\{ \Lambda^2 - M^2 \ln \left(\Lambda^2/M^2 \right) \right\}.$$

This is quadratically divergent, but observe that there is also a logarithmically divergent part. That is a general feature, given a divergent diagram one finds a leading divergence (here quadratic) and also all lower divergencies. These lower divergencies appear usually with and without logarithms. At k loops one will find in general logarithms up to the kth power. For example if some diagram is quadratically divergent at three loops one will obtain terms behaving as Λ^2, $\ln \Lambda$, $\ln^2 \Lambda$ and $\ln^3 \Lambda$.

It is worthwhile to check the dimensionality of the result obtained. In this case the result must be of the dimensionality $[\text{MeV}]^2$ if everything is expressed in MeV. Obviously Λ has the dimension [MeV]. Terms that have an M^2 have already dimensionality [MeV]^2 and contain therefore no Λ^2. Generally a logarithm must be taken as dimensionless, and the argument of the logarithm must be dimensionless also.

It is instructive to work out I_2 with dimensional regularization. Putting $A = M^2$ and $\alpha = 1$ in the previously obtained equation, setting $n = 4 + d$ and using

$$\Gamma\left(-1 - \tfrac{d}{2}\right) = -\frac{\Gamma\left(-\tfrac{d}{2}\right)}{\left(1 + \tfrac{d}{2}\right)} \simeq \frac{2}{d} - 1 + \gamma.$$

one obtains

$$I_2(Q) = i\pi^2 \left\{ M^2 \left(\frac{2}{d} + \ln \pi + \gamma - 1\right) + M^2 \ln M^2 \right\}.$$

While we see again the combination $2/d + \ln \pi + \gamma - 1$ to correspond to $-\ln \Lambda^2$, no counterpart exists to the Λ^2 term. This is a general feature of dimensional regularization. Seemingly divergent parts are zero. For example consider

$$I_6 = \int d_4 q \cdot \left[q^2 + M^2 - i\epsilon\right].$$

No Wick rotation can be applied because the integrals over the segments diverge. It is not needed anyway. The integral obviously behaves like Λ^6. With dimensional regularization, setting $\alpha = -1$, and $n = 4 + d$ one finds:

$$\frac{i\pi^{2 + \frac{d}{2}}}{A^{-3 - \frac{d}{2}}} \cdot \frac{\Gamma(-3 - \frac{d}{2})}{\Gamma(-1)}$$

and this is zero for small d because $\Gamma(-1) = \infty$.

6.3 Self Energy

Let us now go back to the starting point, $I(Q)$. The infinite part is given by

$$-\frac{2i\pi^2}{d}, \quad d \to 0.$$

Thus, limiting ourselves to the infinite part, the result for the diagram with σ-exchange, one loop self-energy insertion, is:

$$\delta_m \cdot \frac{1}{(Q^2 + m^2 - i\epsilon)^2}, \quad \delta_m = -\frac{i\pi^2}{d} \frac{4g^2}{(2\pi)^4 i}.$$

The diagram without self-energy insertion has only one σ-propagator, and combining the two diagrams we have:

$$\frac{1}{Q^2 + m^2 - i\epsilon} + \frac{1}{Q^2 + m^2 - i\epsilon} \delta_m \frac{1}{Q^2 + m^2 - i\epsilon}.$$

We omitted factors common to both diagrams. Now δ_m is of order g^2. Up to terms of order g^4 this is equal to

$$\frac{1}{Q^2 + m^2 - \delta_m - i\epsilon}.$$

It is somewhat bizarre to consider something of the form $g^2\cdot$ infinity as to be of order g^2, but that is what we do. The (infinite) quantity δ_m appears combined with the parameter m^2.

The renormalization idea is the following. Experimentally one observes never the lowest order terms alone, but the sum of all orders. Up to this order, g^4, the mass squared in the propagator is $m^2 - \delta_m$, and that is what the experimenter observes. In the example considered he/she sees it as the range of the Yukawa potential. In other words, $m^2 - \delta_m$ is the observed mass squared. The theory however makes no prediction about the mass. It is a free parameter, and it must be fixed by comparing the results of the theory with the observed data. In other words, m must be chosen such that $m^2 - \delta_m$ is the observed mass squared.

An infinity such as δ_m is basically unobservable if it occurs in combination with a free parameter. For this it is essential that δ_m is a constant and not a function of Q.

It is embarrassing that δ_m is infinite, but we must live with that.

A short note here: suppose that there also would have been an infinity multiplied by Q^2, i.e. suppose that the infinite part of the result is of the form $\delta_m + Q^2\delta_w$, where δ_w is another infinity containing a factor $1/d$. Then the final result would have been

$$\frac{1}{Q^2 + m^2 - \delta_m - Q^2\delta_w - i\epsilon} \sim$$
$$\frac{1}{1 - \delta_w} \cdot \frac{1}{Q^2 + m^2 - \delta_m + m^2\delta_w - i\epsilon},$$

neglecting terms of order g^4 such as $\delta_m\delta_w$ or δ_w^2. Now a propagator always connects two vertices, and the factor $1/(1 - \delta_w)$ can be absorbed in the coupling constant(s) associated with these

vertices. Thus constant infinities as well as infinities with a factor momentum squared can be absorbed in the parameters of the
theory. However, infinities multiplied with other Q dependence,
such as for example $\ln Q^2$ can not be absorbed.

The most important question is the following: do all infinities of the theory appear only
in combination with a few free parameters? If this is the case
we call the theory renormalizable, else non-renormalizable. As
it happens, the simple σ–π theory considered here has only four
infinities, although they can occur repeatedly. For example, one
may have two self-energy insertions, see the figure.

Diagrams like this, that are really repeated diagrams of lower
order, are called reducible. Diagrams that can not be split in two
by cutting just one line are called irreducible.

One may have many such insertions, and all together one has
a series of diagrams:

The corresponding expression is:

$$\frac{1}{Q^2 + m^2 - i\epsilon} \cdot \left[1 + x + x^2 + x^3 + \dots\right], \quad x = \frac{\delta_m}{Q^2 + m^2 - i\epsilon}$$

The whole series may be summed up:

$$= \frac{1}{Q^2 + m^2 - i\epsilon} \cdot \frac{1}{1 - x} = \frac{1}{Q^2 + m^2 - \delta_m - i\epsilon}.$$

By accident the complete result for the infinite part, to all orders
in perturbation theory, equals the previously obtained result to
order g^4.

Another divergent diagram is the π self-
energy diagram. Also the infinite part of
this diagram may be absorbed into a free
parameter, namely the pion mass M.

Finally there are tadpole diagrams, which we will consider further down.

We will now proceed to show that the two diagrams considered
(and the tadpole diagrams) are the only divergent ones.

6.4 Power Counting

The two self-energy diagrams considered in the previous section were logarithmically divergent. Let us now consider a diagram with three external lines. The corresponding expression is:

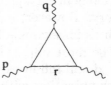

$$\int d_4 r \; \frac{1}{(r^2 + M^2)\left((r+p)^2 + M^2\right)\left((r+p+q)^2 + M^2\right)}.$$

For large r this expression behaves as

$$\int d_4 r \; \frac{1}{r^6},$$

which is convergent.

Compared to the σ self-energy diagram there is one extra propagator. It improves the degree of convergence by two. From this example it is clear that one-loop diagrams with more than two external lines are convergent.

Now a two loop self-energy diagram. There are two four-dimensional integrals:

$$\int d_4 p \int d_4 r \; \frac{1}{(r^2 + M^2)\left((r+q)^2 + M^2\right)} \cdot$$

$$\frac{1}{(p^2 + M^2)\left((p-r)^2 + m^2\right)\left((p+q)^2 + M^2\right)}.$$

For large r and p this behaves as

$$\int d_4 p \, d_4 r \; \frac{1}{r^4 p^4 (p-r)^2},$$

or, as we will write:

$$\int d_8 k \; \frac{1}{k^{10}} \sim \frac{1}{\Lambda^2}.$$

In power counting there is no need to worry about details such as how $p - r$ behaves for large p and r. The integral behaves as $1/\Lambda^2$, i.e., it is convergent.

Exercise 6.4 Consider the double integral

$$\int_a^\Lambda dy \int_a^\Lambda dx \; \frac{1}{xy(x - y + ib)}, \quad a, b > 0.$$

Show that it behaves as $1/\Lambda$ or $\ln(\Lambda)/\Lambda$ for large Λ.

Compared to the one loop self-energy diagram there is one more four-integration, but three more propagators. Thus going from one to two loops one has an extra factor Λ^4/Λ^6 for the purposes of power counting. It is clear that adding loops makes the diagrams more convergent, and we see that indeed the only divergent diagrams are the one loop σ and π self-energy diagrams. That is, apart from tadpole diagrams.

A tadpole diagram is a diagram with one external line. There is one such one loop diagram. The corresponding expression has only one propagator and a combinatorial factor $1/2$:

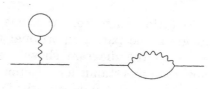

$$\tfrac{1}{2} \int d_4 r \frac{2g}{r^2 + M^2 - i\epsilon}.$$

It is quadratically divergent. To order g^2 this diagram occurs only in combination with the π self-energy diagram. The momentum flowing through the σ line of the first diagram is zero. That diagram gives:

$$\frac{g^2}{m^2 - i\epsilon} \int d_4 r \frac{1}{r^2 + M^2 - i\epsilon} \sim \frac{g^2 \Lambda^2}{m^2}.$$

Also this can simply be absorbed in the π mass M, together with the logarithmic infinity of the second diagram.

On the two loop level the tadpole diagram is still logarithmically divergent. The corresponding integral behaves as

$$\int d_8 k \frac{1}{k^8}.$$

Also this can be absorbed in the π mass.

A very brief example of what happens for two loop diagrams is perhaps instructive. Consider a two loop π self-energy diagram, see figure. It is of order g^4. One recognizes the previously calculated σ self-energy diagram as part of this graph. How must this be handled?

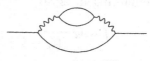

Now the infinity of the one loop σ self-energy diagram was absorbed in the σ mass m. That means that the σ mass is now given by $m^2 + \delta_m$. With this choice for the σ mass however there is a contribution of order g^4 hidden in the one loop diagram, see figure, where the dot signifies that for the σ mass one must take $m^2 + \delta_m$. To order g^2 one has:

$$\frac{1}{Q^2 + m^2 + \delta_m - i\epsilon} = \frac{1}{Q^2 + m^2 - i\epsilon}$$
$$- \frac{1}{Q^2 + m^2 - i\epsilon} \delta_m \frac{1}{Q^2 + m^2 - i\epsilon}$$

which can be pictured as shown.

Selecting then from this one loop diagram the part proportional to g^4 leads to the diagram shown, where now the cross stands for a factor $-\delta_m$.

In other words, having absorbed the one loop infinity in the σ mass means that at the g^4 level we find in fact two diagrams:

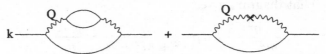

The cross equals precisely minus the infinite part of the σ self-energy insertion in the first diagram. Together then they are finite; in fact, the self-energy diagram was computed before, and together the result is finite and of the form

$$F(Q) = C + \int_0^1 dx \, \ln\left(x(1-x)Q^2 + M^2\right)$$

where C is a constant. For large Q that behaves as $\ln Q^2$. The remaining integral

$$\int d_4Q \, \frac{F(Q)}{\left((Q-k)^2 + M^2\right)\left(Q^2 + m^2\right)^2} \sim \int d_4Q \, \frac{\ln Q^2}{Q^6}$$

is convergent.

In other theories, such as for example quantum electrodynamics, this is not that simple. The remaining integral is usually divergent, and one must show that the remaining infinity is such that it can be absorbed again in the parameters of the theory, i.e., it must be a linear combination of a constant and a constant times momentum squared but nothing else.

Disentangling infinities is a rather complicated affair. One speaks of overlapping divergencies.

6.5 Quantum Electrodynamics

The situation in quantum electrodynamics is more complicated. The reason is that the electron propagator behaves as $1/k$ for large momentum rather than $1/k^2$ as the π and σ propagators in the foregoing. The divergent diagrams at the one loop level are:

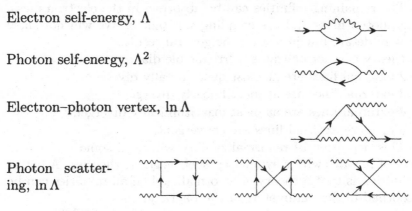

Electron self-energy, Λ

Photon self-energy, Λ^2

Electron–photon vertex, $\ln \Lambda$

Photon scattering, $\ln \Lambda$

Exercise 6.5 Verify this.

The tadpole diagram is zero. The corresponding expression is:

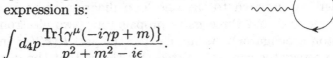

$$\int d_4p \frac{\text{Tr}\{\gamma^\mu(-i\gamma p + m)\}}{p^2 + m^2 - i\epsilon}.$$

The trace gives $-4ip_\mu$, the resulting integral is:

$$\int d_4p \frac{-4ip_\mu}{p^2 + m^2 - i\epsilon}.$$

All p-integrations are symmetric and go from $-\infty$ to $+\infty$, but the integrand is odd in p and the integral is therefore zero.

Exercise 6.6 Show that

$$\int_{-\infty}^{+\infty} dx \; x \; f(x^2) = 0, \qquad \int_{-\infty}^{1+\infty} dx \; x \; f(x^2) \neq 0.$$

One speaks of symmetrical integration, the part for positive p_μ cancels the part for negative p_μ. When using any regulator method one must take care not to destroy this cancellation, or else all kinds of disasters result. In particular translations such as $p_\mu \to p_\mu + q_\mu$ must be treated with care.

It may be shown that the sum of all photon–photon scattering diagrams is finite as well. Also that depends on how one treats the infinities, but all those details will not be discussed here, although they are very important.

The remaining infinities can be absorbed in the electron mass, the photon mass and the coupling constant e. We will not show that in detail, but just state the general result:
A theory is renormalizable if irreducible diagrams with

2 external lines are at most quadratically divergent;

3 external lines are at most linearly divergent;

4 external lines are at most logarithmically divergent;

5 or more external lines are convergent.

This definition of renormalizability will need some revision in case of theories with a symmetry, such as gauge theories. A better definition is that a theory is renormalizable if all infinities can be absorbed in the parameters of the Lagrangian.

Let us now consider what happens for two loops. Consider a two loop electron self-energy diagram, as  shown. Compared to the one loop diagram there is one more four-integral and three more propagators (two electron and one photon propagator). Assuming that the part $k_\mu k_\nu / \kappa^2$ in the photon propagator can be omitted this shows that the degree of divergence remains unchanged compared to the one loop case. This is generally true. Thus higher order diagrams give the same type of infinities as the one loop diagrams. Also those infinities can be absorbed in the free parameters of the theory.

It is crucial for renormalizability that the $k_\mu k_\nu / \kappa^2$ part of the photon propagator can be omitted. The proof that this can be done requires consideration of Ward identities, a consequence of gauge invariance.

The renormalization procedure for a case like quantum electrodynamics is an order by order procedure. First one considers one loop diagrams and absorbs all infinities in the free parameters. Next consider two loop diagrams. New infinities arise, of higher order in the coupling constant, also to be absorbed; and so on. A complication is that one must show that the new infinities can indeed be absorbed. For example, an infinity to be absorbed in a mass must not be momentum dependent.

Exercise 6.7 Consider the simple π–σ model. Add a new interaction, with five lines, four π and one σ line. The corresponding interaction Hamiltonian is $\mathcal{H} = g\sigma\pi^4$. Show that there is a one loop quadratically divergent σ self-energy diagram. Show that there is a quartic divergent three loop σ self-energy diagram. Show that there is a one loop diagram with two external σ and four external π lines that is logarithmically divergent. Try to analyse what happens in general. The theory is non-renormalizable.

6.6 Renormalizable Theories

On the basis of the foregoing it is quite simple to make a list of all possible renormalizable theories. Denoting a scalar field by ϕ and a fermion field by ψ, the allowed interactions are of the form:

$$\phi^3$$
$$\phi^2\partial\phi$$
$$\phi^4$$
$$\phi(\bar{\psi}\psi)$$

A derivative in a vertex as shown above increases the degree of divergence of the theory as it adds one momentum for every such vertex occurring. One such derivative can be tolerated in a three-vertex. With three scalar fields such a vertex can actually not occur as it is not Lorentz invariant, but the possibility arises with vector fields.

If in a vector field propagator the part $k_\mu k_\nu / M^2$ can be omitted then a vector field may be considered as a scalar field for the purposes of power counting. Here are some examples of renormalizable interactions:

$$V_\nu V_\mu \partial_\mu V_\nu$$
$$V_\mu \phi \partial_\mu \phi$$
$$V_\mu (\bar{\psi} \gamma^\mu (a + b\gamma^5)\psi)$$
$$\phi(\bar{\psi}(a + b\gamma^5)\psi)$$
$$V_\mu V_\mu V_\nu V_\nu$$

All of these vertices occur in the Standard Model.

For theories with vector fields the crucial question is the behaviour of the vector field propagator. In the case of gauge theories without anomalies that turns out to be all right, but the analysis requires a considerable development of the method.

6.7 Radiative Corrections: Lamb Shift

In practice, computing one loop diagrams, renormalization is a rather trivial procedure. The infinities are there, but the quantities of practical interest are finite. For example, radiative corrections to vector boson masses in the Standard Model are infinite, but the radiative corrections to the ratio of charged to neutral vector boson mass (including the cosine of the weak mixing angle as a factor) are finite. Other examples are the calculation of $g - 2$ of muon and electron (the anomalous magnetic moment), and the Lamb shift. At the two loop level things are much more complicated: first all infinities at the one loop level must be computed, they must be subtracted, and then the two loop calculation can be done. Only then must the whole renormalization formalism be applied.

Let us start with the Lamb shift. An important contribution is due to vacuum polarization, that is self-energy of the photon. That implies a modification of the photon propagator, and we will compute that piece.

As we have seen, the Coulomb potential is really nothing else but the Fourier transform of the photon propagator in the no-recoil approximation:

$$\frac{1}{(2\pi)^4 i} \frac{1}{4\pi} \int d_3 r \, \frac{e^{i\vec{Q}\vec{r}}}{r} = \frac{1}{(2\pi)^4 i} \frac{1}{Q^2}$$

where $Q = |\vec{Q}|$ is the magnitude of the three-momentum $\vec{Q}$ passing through the photon propagator. Radiative corrections modify the photon propagator, and perforce its Fourier transform. Thus there is a slight modification to the $1/r$ potential.

The relevant one loop photon self-energy diagram is as shown in the figure. The fermion closed loop may involve electrons, muons, quarks, etc., but as we will see, only the lowest mass particle (the electron) gives a significant contribution. The corresponding expression is:

$$\Pi_{\mu\nu}(Q) = -(ie)^2 \int d_4q \frac{\text{Tr}\left[\gamma^\mu(-i\gamma q + m)\gamma^\nu(-i\gamma(q + Q) + m)\right]}{(q^2 + m^2)((q + Q)^2 + m^2)}.$$

Writing down such an expression in case of fermions involves the rule: read against the arrow. Thus start at the left vertex, get the expression for the lower line, then the right vertex, then the upper line. Note the minus sign for the fermion loop. As a matter of notation we write γq for $\gamma^\alpha q_\alpha$. The calculation proceeds in several steps. First the trace is done, with the result:

$$\text{Tr}\left[\ldots\right] = -8q_\mu q_\nu - 4q_\mu Q_\nu - 4Q_\mu q_\nu + 4\delta_{\mu\nu}(m^2 + q^2 + qQ).$$

As usual, qQ denotes the dot-product between q and Q. Next the integral must be done. Introducing a Feynman parameter x the integral becomes:

$$\int_0^1 dx \int d_4q \frac{\text{Tr}\left[\ldots\right]}{(q^2 + 2x\,qQ + xQ^2 + m^2)^2}.$$

Now introduce the new integration variable p instead of q, with $q = p - x\,Q$. The denominator of the integrand becomes:

$$\left(p^2 + M^2\right)^2, \quad \text{with} \quad M^2 = m^2 + x(1 - x)Q^2.$$

Of course, the substitution $q = p - x\,Q$ must also be done in the result for the trace. Since the denominator is even in the integration variable p all terms odd in p may be dropped (symmetrical integration). The result is:

$$\Pi_{\mu\nu}(Q) = e^2 \int dx \int d_4p$$

$$\frac{-8p_\mu p_\nu + 8x(1 - x)Q_\mu Q_\nu + 4\delta_{\mu\nu}(m^2 - x(1 - x)Q^2 + p^2)}{(p^2 + M^2)^2}$$

The integration over the momentum p is done using dimensional regularization. Three different integrals need to be done, they are (see appendix on dimensional regularization):

$$\int d_4p \, \frac{p_\mu p_\nu}{(p^2 + M^2)^2} = i\pi^2 \delta_{\mu\nu} \left[\frac{M^2}{n-4} - \tfrac{1}{2} M^2 + \tfrac{1}{2} M^2 \ln(M^2) \right]$$

$$\int d_4p \, \frac{p^2}{(p^2 + M^2)^2} = i\pi^2 \left[\frac{4M^2}{n-4} - M^2 + 2M^2 \ln(M^2) \right]$$

$$\int d_4p \, \frac{1}{(p^2 + M^2)^2} = i\pi^2 \left[\frac{-2}{n-4} - \ln(M^2) \right]$$

The result will contain terms with $\ln(M^2)$. With $M^2 = m^2 + x(1-x)Q^2$ and assuming $Q^2 \ll m^2$ we may expand, $\ln(M^2) = \ln(m^2) + x(1-x)Q^2/m^2$. The remaining integral over x becomes elementary, as only terms involving some power of x occur. The final result is:

$$\Pi_{\mu\nu}(Q) = i\pi^2 e^2 \left(Q_\mu Q_\nu - Q^2 \delta_{\mu\nu} \right)$$
$$\cdot \left[-\tfrac{8}{3} \frac{1}{n-4} - \tfrac{4}{3} \ln(m^2) - \tfrac{4}{15} \frac{Q^2}{m^2} \right].$$

Consider now the corrected photon propagator. Adding in this radiative correction 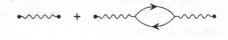 one has the two diagrams shown. In lowest order there is only the first diagram, the photon propagator, whose Fourier transform was given above. The second diagram containing two photon propagators must be added, that is we must add

$$\frac{1}{(2\pi)^4 i} \frac{1}{Q^2} \Pi_{\mu\nu}(Q) \frac{1}{(2\pi)^4 i} \frac{1}{Q^2}.$$

As observed several times before, the terms $Q_\mu Q_\nu$ in this expression may be ignored (they give zero when multiplied into the vertices that connect to the propagator). The total is now:

$$\frac{1}{(2\pi)^4 i} \frac{1}{Q^2} \left[1 + C + \lambda Q^2 \right] \delta_{\mu\nu}$$

with

$$C = \frac{i\pi^2 e^2}{(2\pi)^4 i} \left[\tfrac{8}{3} \frac{1}{n-4} + \tfrac{4}{3} \ln(m^2) \right]$$

$$\lambda Q^2 = \frac{i\pi^2 e^2}{(2\pi)^4 i}\left[\frac{4Q^2}{15m^2}\right] = \frac{\alpha}{15\pi}\frac{Q^2}{m^2}.$$

As usual, $\alpha = e^2/4\pi \approx 1/137$.

The term C just amounts to a modification of the coefficient of the lowest order result. It gives the same $1/r$ behaviour, only the coefficient is affected. This is then charge renormalization, the initial e must be chosen such that the combination $e^2(1+C)$ is the experimentally observed charge. The interesting piece is the second part. That adds a finite term with different behaviour with respect to r. As a function of Q it behaves as a constant. Since obviously

$$\int d_3 r\, e^{i\vec{Q}\vec{r}}\delta_3(\vec{r}) = 1,$$

we see that this term amounts then to a change

$$\frac{1}{r} \quad \rightarrow \quad \frac{1}{r} + \frac{4\alpha}{15m^2}\delta_3(\vec{r}).$$

This corrected potential has a new piece that is effective at the origin ($\vec{r}=0$) only. It will affect wave functions that are non-zero at the origin, which for the hydrogen atom are S-waves only. Therefore only S-waves are affected, causing an energy split between the $2s\frac{1}{2}$ and $2p\frac{1}{2}$ states. That then is part of the corrections that are observed in the Lamb shift. Other contributions to the Lamb shift result from radiative corrections due to triangle diagrams and electron self-energy diagrams.

6.8 Radiative Corrections: Top Correction to ρ-Parameter

Another example of radiative corrections needing calculation of one loop self-energy diagrams only is the correction to the ρ-parameter. This parameter is about the ratio of charged and neutral vector boson masses. In the present day formulation of the Standard Model the vector boson masses arise through the Higgs mechanism, and assuming the simplest Higgs system there is actually a relation between charged and neutral vector boson masses. This relation is expressed through a parameter which is in that simplest case equal to one:

$$\rho = \frac{M^2}{c_w^2 M_0^2} = 1 + \text{radiative corrections} \quad \text{(if simplest Higgs).}$$

In this, $M \approx 80.25$ and $M_0 \approx 91.17$ GeV are the charged and neutral vector boson masses respectively. Further $c_w^2 \approx 0.7663$ is the square of the cosine of the weak mixing angle ($s_w^2 = \sin^2 \theta_w \approx 0.2337$). By far the largest error margins are those on M, namely about 0.3%, i.e., 0.6% on ρ. Using the numbers cited we have, experimentally, $\rho = 1.011 \pm 0.006$. In first approximation the radiative correction due to a large top mass equals $+3G_F m_t^2 / 8\pi^2 \sqrt{2} \approx 0.0031 m_t^2$, with m_t in GeV. From this one finds $m_t \approx 180$ GeV; a more careful analysis including all kinds of other radiative corrections yields $m_t = 150(+23, -28)$ GeV*. We must emphasize that the experimental result for ρ strongly suggests that the Higgs system realized in the Standard Model is the simplest one. This is the truly important conclusion. But let us now derive the equation for the leading correction proportional to the top mass squared.

The interactions needed to derive the desired equation are the interactions of the charged (W) and neutral (Z) vector bosons with top and bottom quark. The relevant interaction Lagrangian is (ignoring Cabibbo–Kobayashi–Maskawa complications):

$$\mathcal{L} = \frac{ig}{2\sqrt{2}} \left[W_\mu^+ \left(\bar{t} \gamma^\mu (1 + \gamma^5) b \right) + W_\mu^- \left(\bar{b} \gamma^\mu (1 + \gamma^5) t \right) \right]$$

$$+ \frac{ig}{4c_w} \left[Z_\mu \left(\bar{t} \gamma^\mu (1 - \tfrac{8}{3} s_w^2 + \gamma^5) t \right) + Z_\mu \left(\bar{b} \gamma^\mu (\tfrac{4}{3} s_w^2 - 1 - \gamma^5) b \right) \right].$$

The W_μ^+ field can annihilate a W^+ and create a W^-, the complex conjugate field W_μ^- can annihilate a W^- and create a W^+. The Z_μ field can annihilate and create a Z^0. We have not indicated that there are in fact three different top and bottom quarks; we refer to colour. Quarks of different colour have the same interaction with the vector bosons. At the end we must sum over possible colour states.

We start by computing the diagram shown in the figure. As vertex factor we take $i\gamma^\mu(a + b\gamma^5)$ for the left vertex and $i\gamma^\mu(a' + b'\gamma^5)$ for

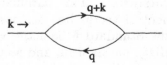

---

* Numerical values taken from Particle Properties 1992.

the right vertex, specifying later the precise values for the actual cases. The corresponding expression is:

$$S_{\mu\nu}(a, b, a', b', m, m') = \int d_n q$$

$$\frac{[\gamma^\nu(a + b\gamma^5)(-i\gamma q + m)\gamma^\mu(a' + b'\gamma^5)(-i\gamma(q + k) + m')]}{(q^2 + m^2 - i\epsilon)((q + k)^2 + m'^2 - i\epsilon)}.$$

This includes a minus sign for the fermion loop and a minus from the square of the i in the vertex. The trace must be taken of the expression in square brackets. We compute the contribution of this type of self-energy graphs at zero momentum, that is as they appear at low energies, in muon decay and neutrino–electron scattering. Since the ρ-parameter, and in any case the contribution proportional to the top mass squared, depend only slightly on energy, that is sufficient for our purposes. Thus we set $k = 0$.

The trace is now readily done:

$$[] = 4(aa' + bb')(q^2\delta_{\mu\nu} - 2q_\mu q_\nu) + 4m'm(aa' - bb')\delta_{\mu\nu}.$$

The equations given in the appendix on dimensional regularization are entirely adequate to deal with this situation. Using

$$q^2 = \tfrac{1}{2}(q^2 + m^2) + \tfrac{1}{2}(q^2 + m'^2) - \tfrac{1}{2}(m^2 + m'^2)$$

one obtains, as a matter of notation:

$$S_{\mu\nu}(a, b, a', b', m, m') = 2(aa' + bb')\,\delta_{\mu\nu}\,(A(m')+$$
$$A(m) - 4B_{22}(0, m, m'))$$
$$- 2(aa' + bb')\,\delta_{\mu\nu}(m'^2 + m^2)\,B_0(0, m, m')$$
$$+ 4(aa' - bb')\,\delta_{\mu\nu}m'm\,B_0(0, m, m').$$

The following equations from the appendix are being used:

$$A(m) = im^2\Delta + i\pi^2 m^2 \ln(m^2) - i\pi^2 m^2;$$
$$B_0(0, m, m') = -i\Delta + i\pi^2$$
$$\left(1 + \frac{m^2}{m'^2 - m^2}\ln(m^2) - \frac{m'^2}{m'^2 - m^2}\ln(m'^2)\right);$$

$$B_{22}(0, m, m') = \left(\frac{i\Delta}{4} - \frac{3i\pi^2}{8}\right)(m^2 + m'^2)$$

$$+ \frac{i\pi^2}{4(m'^2 - m^2)}\left(m'^4 \ln(m'^2) - m^4 \ln(m^2)\right) ;$$

$$B_0(0, m, m) = -i\Delta - i\pi^2 \ln(m^2) ;$$

$$B_{22}(0, m, m) = \frac{i\Delta}{2}m^2 - \frac{i\pi^2}{2}m^2 + \frac{i\pi^2}{2}m^2 \ln(m^2) .$$

The quantity Δ contains an infinity, but it will cancel out so we will not bother about it.

We must now recall how a self-energy diagram leads to a mass correction. One can either sum the series of self-energy insertions, or simply check at second order. For a change we will do the latter.

Consider the diagrams shown in the figure. Let $g^2 S\delta_{\mu\nu}$ be the expression for the self-energy diagram, as calculated above. The result is:

$$\frac{1}{(2\pi)^4 i} \frac{\delta_{\alpha\beta}}{k^2 + M^2 - i\epsilon}$$

$$+\frac{1}{(2\pi)^4 i} \frac{\delta_{\alpha\mu}}{k^2 + M^2 - i\epsilon} g^2 S\delta_{\mu\nu} \frac{1}{(2\pi)^4 i} \frac{\delta_{\nu\beta}}{k^2 + M^2 - i\epsilon} .$$

Apart from terms of order g^4 this is equal to:

$$\frac{1}{(2\pi)^4 i} \frac{\delta_{\alpha\beta}}{k^2 + M^2 - g^2 S/(2\pi)^4 i - i\epsilon} .$$

Thus a self-energy diagram value of $g^2 S$ gives a mass squared correction of $-g^2 S/(2\pi)^4 i$.

Consider now the correction to the ρ-parameter. If $\rho = M^2/c_w^2 M_0^2 = 1$ without radiative corrections then we will now obtain:

$$\rho = \frac{M^2 + \delta_+}{c_w^2(M_0^2 + \delta_0)} = 1 + \delta\rho = 1 + \frac{\delta_+ - c_w^2 \delta_0}{M^2} ,$$

where δ_+ and δ_0 are the corrections to M (charged vector boson mass) and M_0 (neutral vector boson mass). It follows

$$\delta\rho = -\frac{1}{M^2(2\pi)^4 i}\left[A^+ - c_w^2 A^0\right]$$

where A^+ and A^0 correspond to the diagrams shown in the figure.

The general self-energy diagram was computed before, and we have:

$$A^+ - c_w^2 A^0 = \left(\frac{g}{2\sqrt{2}}\right)^2 S(1,1,1,1,m_b,m_t)$$

$$- c_w^2 \left(\frac{g}{4c_w}\right)^2 \left(S(a_t,1,a_t,1,m_t,m_t)\right.$$

$$+ S(a_b,-1,a_b,-1,m_b,m_b)) ,$$

where

$$a_t = 1 - \tfrac{8}{3}s_w^2 \quad \text{and} \quad a_b = \tfrac{4}{3}s_w^2 - 1$$

specify the top and bottom couplings to the neutral vector boson. Using the expression for S and the equations for the functions A and B given above, and remembering that there are three top and bottom quarks we obtain:

$$\rho = 1 + \frac{3G_F}{8\pi^2\sqrt{2}}\left[m_t^2 + m_b^2 - \frac{2m_t^2 m_b^2}{m_t^2 - m_b^2}\ln\left(\frac{m_t^2}{m_b^2}\right)\right].$$

We have used the relation between g and the Fermi constant G_F:

$$G_F = \frac{g^2\sqrt{2}}{8M^2},$$

with $G_F = \dfrac{1.0246 \times 10^{-5}}{m_p^2} = 1.1638 \times 10^{-11}\text{ MeV}^{-2}$.

In here $m_p = 938.272$ MeV is the proton mass. It may be noted that sometimes the factor $\sqrt{2}$ is incorporated in the definition of G_F. The numerical value of G_F is from comparison with experiments on muon decay. Being careful with the logarithm it may be verified that the correction is zero for equal masses; write $m_t^2 = m_b^2(1+\delta)$ and find the lowest order result in terms of δ.

The bottom mass $m_b \approx 5$ GeV is small with respect to the top mass, known to be above 90 GeV, and taking the limit $m_b \to 0$ gives the final result:

$$\rho = 1 + \tfrac{3}{8}\frac{G_F m_t^2}{\pi^2\sqrt{2}}.$$

This correction to the ρ-parameter is a very special one as it is at this time the only one known to grow as the mass of the intermediate particle grows. For example, in radiative corrections to things like the Lamb shift and the anomalous magnetic moment of muon and electron one can forget about heavy particles. In a certain sense that is a pity: one has then no window on the mass spectrum at high energies. The very interesting thing about the ρ-parameter is that we not only get the message that the Higgs sector is the most simple one, but also that there are no more top–bottom (or for that matter heavy lepton–neutrino) doublets with large mass differences. We heave a window on very heavy masses. Actually, also mass differences in other multiplets can be excluded. This puts a limit on model building when trying to go beyond the Standard Model.

The (small) deviation from one for the ρ-parameter can be put in terms of a symmetry, sometimes called custodial symmetry. It is actually nothing else but good old isospin symmetry. The top and bottom quark can be considered as an isospin doublet; the mass difference represents breaking of that symmetry. Thus, we learn that for very heavy masses isospin must be respected.

6.9 Neutral Pion Decay and the Anomaly

In this section we will derive the modification to the PCAC equation due to the Adler–Bell–Jackiw anomaly. This means that we must work out the Ward identity that normally leads to PCAC and then find out what goes wrong. To understand the arguments of this section it is necessary to know the material discussed previously on PCAC and pion decays. Let us recall the essentials.

The PCAC equation for the axial current is:

$$\partial_\alpha j_\alpha = \sqrt{2} f_\pi m_\pi^2 \pi.$$

It was "derived" using for the axial current the combination $i(\bar{u}\gamma^\alpha\gamma^5 d)$ involving the anti-up and down quark. The pion field was identified, apart from a factor, with the combination $i(m_u + m_d)(\bar{u}\gamma^5 d)$. That factor is always the same and is in fact the factor in front of the pion field on the right hand side of the PCAC equation. The derivation relied on use of the Dirac equation for the $\bar{u}$ and d quantities, which is not correct if the axial current connects to propagators. It is this latter point that we must investigate.

The process considered is π^0 decay. We first write down the relevant diagrams, then take the divergence (multiplying with the pion momentum) and then try to recover the PCAC equation. That is much the same work as before, except that we have now propagators rather then spinors satisfying the Dirac equation. The technique is fairly general, and is typical for deriving Ward identities. We must deal now with the π^0, which couples to a mixture of $(\bar{u}\gamma^5 u)$ and $(\bar{d}\gamma^5 d)$, in fact exactly the same way the neutral vector boson couples (axial part) to these quarks in the Standard Model, but with zero mixing angle ($c_w = 1$). Compared to the charged current the mixture is

$$\pi^0 \propto \frac{i}{\sqrt{2}} \left[(\bar{u}\gamma^5 u) - (\bar{d}\gamma^5 d) \right] .$$

We first concentrate on the $(\bar{u}\gamma^5 u)$ piece, the other follows then easily. Thereby one must also remember that up and down quarks couple differently to the photon (charges $2/3$ and $-1/3$ respectively). And they all come in three colours.

So, let us start by considering the diagrams for the process at hand, the axial current, $i(\bar{u}\gamma^\alpha\gamma^5 u)$, going to two photons. Note our convention for the momenta: they are all ingoing. It is easy enough to put in the end whatever is desired, and it really pays to be very systematic. There are two diagrams, shown in the figure. The second is obtained by interchanging the two photons. It is also the same as that obtained by reversing the fermion direction in the first diagram. Using the one or the other gives in the end the same, but the starting expression may look different, depending on how momenta are choosen. But neither way is the best for investigating Ward identities. To derive a Ward identity with respect to some vertex (here the axial current vertex) try to put the diagrams such that it appears as a basic diagram with that vertex inserted at all possible places, here two. The figure shows the idea.

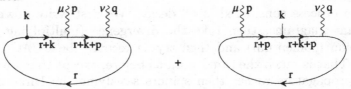

Including the factor $1/\sqrt{2}$ as mentioned in the definition of the π^0 field the expressions corresponding to these two diagrams are:

$$\frac{-i(iec)^2}{(2\pi)^4 i\sqrt{2}} \int d_n r$$

$$\frac{[(-i\gamma(r+k)+m)\gamma^\alpha \gamma^5 (-i\gamma r+m)\gamma^\nu (-i\gamma(r+k+p)+m)\gamma^\mu]}{(r^2+m^2)((r+k)^2+m^2)((r+k+p)^2+m^2)}$$

and

$$\frac{-i(iec)^2}{(2\pi)^4 i\sqrt{2}} \int d_n r$$

$$\frac{[(-i\gamma(r+k+p)+m)\gamma^\alpha \gamma^5 (-i\gamma(r+p)+m)\gamma^\mu (-i\gamma r+m)\gamma^\nu]}{(r^2+m^2)((r+p)^2+m^2)((r+k+p)^2+m^2)} .$$

The square brackets imply a trace. The expressions are read from the graphs by reading against the arrow. In both cases we started with the propagator to the right of the axial vector vertex. The minus sign is because of the fermion loop. There is a remaining factor $1/(2\pi)^4 i$ because no factor $(2\pi)^4 i$ is to be given to the axial vector vertex. Remember, we did not give such factor in the previous treatment either. We did not write the overall δ-function for conservation of momentum $(k+p+q=0)$. The charge of the quark in terms of the unit charge e is denoted by c.

We must now take the four-divergence of this expression, i.e., multiply by ik_α. Initially we do as before, writing k in terms of the momentum of the left line minus that of the right line, and then make Dirac equation-like expressions. For the first term:

$$i\gamma k\gamma^5 = (i\gamma(r+k)+m)\gamma^5 + (-i\gamma r+m)\gamma^5 - 2m\gamma^5$$
$$= (i\gamma(r+k)+m)\gamma^5 + \gamma^5(i\gamma r+m) - 2m\gamma^5 + 2i\gamma\tilde{r}\gamma^5 .$$

The last step is really the central point in relation to the anomaly. We anticommuted γr to the right, through γ^5. But γ^5 is defined, also if there are more than four dimensions, as the product

$\gamma^1\gamma^2\gamma^3\gamma^4$, just as in four dimensions. There is simply no other consistent way of handling γ^5. It gives rise to awkward things, and calculations in the Standard Model on the two or more loop level are encumbered by that. Anyway, the vector r is the integration vector, and is as such not limited to four dimensions. All other vectors are, and the range of the indices α, μ and ν is from one to four.

Before we go on, try to not be confused between γ^5 as discussed above and the γ-matrix associated with the fifth dimension. We will try to avoid mentioning that one explicitly.

Now, think of the n-dimensional space as 6, or 8, or 10 or whatever dimensions. Then there will be non-zero components of r beyond the first four. But while γ^1–γ^4 anticommute with γ^5, the matrices γ^λ with $\lambda > 4$ commute. Therefore we have the rule:

$$\gamma r \gamma^5 = -\gamma^5 \gamma r + 2\gamma \tilde{r} \gamma^5 \,.$$

The vector $\tilde{r}$ is the vector r with the first four components removed. As we take the limit to four dimensions the vector $\tilde{r}$ tends to zero, and that will give a factor $n - 4$. But because the integral is divergent there is a pole term, $1/(n-4)$. And together a finite remnant is left over. That is the anomaly. But before we see all this explicitly we must first treat the other terms in the last identity.

In the simple case where there were spinors left and right of the axial vector vertex the first two terms on the right hand side gave zero, the third term (with coefficient $2m$) is the one that we identified as the pion field, and the last term was not there. Basically, we must now show that this last term gives the contribution hinted at before. However we still must show that the first two terms vanish.

The Dirac equation-like terms would give zero if they apply to a spinor, but applied to a propagator they give 1. In other words, they are the inverse of the propagators. For example:

$$\frac{(-i\gamma(r+k)+m)}{(r+k)^2+m^2}\,(i\gamma(r+k)+m) = \frac{(r+k)^2+m^2}{(r+k)^2+m^2} = 1\,.$$

There are two terms, one cancelling the left propagator, the other the propagator on the right hand side. Thus we obtain two terms each with a missing propagator. The same holds for the expression corresponding to the second diagram. With respect to these Dirac equation-like terms we can depict the situation then as in the figure.

These diagrams add up to zero. In particular, the second diagram cancels against the third; the only difference is that the k-line in the second diagram is on the left of the p-line, while in the third diagram it is on the right. That means, in fact, that in the second diagram we have $\gamma^5\gamma^\mu$ and in the third $\gamma^\mu\gamma^5$. But as μ is physical (in the range 1–4) we know γ^μ anticommutes with γ^5, and the two terms add up to zero.

Similarly the first and last term; however, one must do one extra thing, namely make a shift in the integration variable r to $r + k$ in the last diagram before they become entirely the same.

That is typical for the way Ward identities work. One obtains these two terms that eliminate propagators. And then the term with the right hand side propagator eliminated cancels against the next term with the left hand side propagator eliminated.

Thus, in the end, the Dirac equation type terms still cancel, in some sense the Dirac equation still works.

Finally we must evaluate the anomaly. The expression to be evaluated is:

$$\frac{-i(iec)^2 2i}{(2\pi)^4 i\sqrt{2}} \int d_n r$$

$$\frac{\mathrm{Tr}\left[(-i\gamma(r+k)+m)\gamma\tilde{r}\gamma^5(-i\gamma r+m)\gamma^\nu(-i\gamma(r+k+p)+m)\gamma^\mu\right]}{(r^2+m^2)((r+k)^2+m^2)((r+k+p)^2+m^2)}.$$

Let us rewrite this a bit to facilitate evaluation of the trace. We replace $k + p$ in the last propagator by $-q$, and then rotate the trace such that γ^5 is at the end. Now to have something non-zero one needs at least four γ's in the range 1–4, i.e., the factor $\gamma\tilde{r}$ is no good for that. But that last factor itself must combine with some other γ, in the > 4 range, and we conclude that we need all the γ's that we can get. In other words, the mass terms may be dropped. Finally we write all r as $\underline{r} + \tilde{r}$, where $\underline{r}$ is a vector with the first four components of r, but no components beyond that.

The result is now:

$$\frac{-i(iec)^2 2i(-i)^3}{(2\pi)^4 i\sqrt{2}} \int d_n r$$

$$\frac{\text{Tr}\left[(\gamma\underline{r} + \gamma\tilde{r})\gamma^\nu(\gamma\underline{r} + \gamma\tilde{r} - \gamma q)\gamma^\mu(\gamma\underline{r} + \gamma\tilde{r} + \gamma k)\gamma\tilde{r}\gamma^5\right]}{(r^2 + m^2)((r + k)^2 + m^2)((r - q)^2 + m^2)} \cdot$$

At this point there is not much else to do but to work it out. There are 18 terms, but the only non-zero terms are those that have two factors $\gamma\tilde{r}$. Note that these factors anticommute with all other factors, and the product of two such factors, $\gamma\tilde{r}\gamma\tilde{r}$, is simply the dot-product $(\tilde{r}\tilde{r})$. This eliminates all $\gamma\tilde{r}$. What is left are terms with four γ's and a γ^5. The trace of these terms will give an ϵ tensor, and four different vectors must be associated with these vectors. Throwing out all terms that contain two $\underline{r}$ leaves us with 5 terms, of which 4 contain one $\gamma\underline{r}$. Anticommuting this $\gamma\underline{r}$ to the right (as only terms with four γ's survive we need not bother about the δ terms in the anti-commutation rule) makes it clear that those four terms cancel, and we are left with one term:

$$-(\tilde{r}\tilde{r})\text{Tr}\left[\gamma^\nu\gamma q\gamma^\mu\gamma k\gamma^5\right] .$$

This trace is now trivial. Writing $k = -p - q$, of which then the q term gives zero, we have the result

$$\frac{-8e^2 c^2}{(2\pi)^4\sqrt{2}}\epsilon_{\nu\lambda\mu\kappa}q_\lambda p_\kappa \int d_n r$$

$$\frac{(\tilde{r}\tilde{r})}{(r^2 + m^2)((r + k)^2 + m^2)((r - q)^2 + m^2)} \cdot$$

The integral looks formidable but is not. Since the dot-product $(\tilde{r}\tilde{r})$ will give something proportional to $n - 4$ we are really only interested in the divergent part of this integral. That means that we can forget about all momentum dependence other than r in the denominator. We evaluate

$$\int d_n r \, \frac{r_\mu r_\nu}{(r^2 + m^2)^3} \cdot$$

and multiply then with $\delta_{\mu\nu}$ summing μ and ν over the range 4 to n. The integral is given in the appendix on dimensional regularization as $\delta_{\mu\nu}G_3$, and taking only the infinite part gives:

$$\int d_n r \, \frac{r_\mu r_\nu}{(r^2 + m^2)^3} \to -\frac{2i\pi^2}{4(n - 4)}\delta_{\mu\nu} .$$

Now multiplying this with $\delta_{\mu\nu}$ and summing over the range 4 to n simply gives $n-4$. The result is:

$$\frac{4i\pi^2 e^2 c^2}{(2\pi)^4\sqrt{2}}\epsilon_{\nu\lambda\mu\kappa}q_\lambda p_\kappa.$$

To this we must add the term coming from the second diagram, to be obtained by exchanging μ and ν and at the same time p and q. That gives the same expression, and our result for the up quark, one colour, contribution is

$$\frac{8i\pi^2 e^2 c^2}{(2\pi)^4\sqrt{2}}\epsilon_{\nu\lambda\mu\kappa}q_\lambda p_\kappa = \frac{i\sqrt{2}\,\alpha}{\pi}c^2\,\epsilon_{\nu\lambda\mu\kappa}q_\lambda p_\kappa,$$

where $\alpha = e^2/4\pi = 1/137$ is the fine structure constant. Consider now the factor c. The charge of the up quark is 2/3, thus $c^2 = 4/9$, and next summing over the three colours gives 4/3. The contribution of the down quark, charge $-1/3$, is 1/3, and must be subtracted (remember the combination $(\bar{u}u) - (\bar{d}d)$ for the π^0). The final result is as asserted before:

$$\frac{i\sqrt{2}\,\alpha}{\pi}\epsilon_{\nu\lambda\mu\kappa}q_\lambda p_\kappa.$$

which translates to the modified PCAC hypothesis for the neutral axial current:

$$\partial_\alpha j_\alpha^0 = \sqrt{2}f_\pi m_\pi^2 \pi^0 + \frac{i\sqrt{2}\,\alpha}{8\pi}\epsilon_{\nu\mu\alpha\beta}F_{\mu\nu}F_{\alpha\beta}.$$

7

Massive and Massless Vector Fields

7.1 Subsidiary Condition Massive Vector Fields

The electromagnetic field, classically, is a vector field. It is however not obvious that therefore the quantum field, constructed as we did, must be a vector field. In the end that comes out anyway, but let us make the situation clear.

We can freely use scalar fields, vector fields, spinor fields, as long as the theory gives rise to results agreeing with the observed data. One of the required properties is Lorentz invariance, and we must take care that Lorentz invariance is maintained. Our classification in terms of the fields mentioned is really done that way to keep this invariance transparent. We must know precisely how the fields behave under Lorentz transformations, and then we can make them interact such that the invariance is maintained.

The choice as to what kind of field describes an observed particle is really a matter of choice: try what type of field describes best the observed data. So, to anyone criticising the use of a vector field to describe photons one can simply answer: this works well.

Part of the observed phenomena is that there are two kinds of photons, thus two degrees of freedom (polarizations). The photon is presumably massless. The massive vector bosons of weak interactions have three degrees of freedom. The problem arises to construct a set of fields such that there are two (three) degrees of freedom, and such that the fields transform under Lorentz transformations in the proper way. If we take it that the electromagnetic field is described by a vector field (i.e., a set of four fields that under a Lorentz transformation transform into each other according to the Lorentz transformations themselves) then we have a problem, because a vector field has four components, meaning that we have four degrees of freedom. Somehow there must be a

169

restriction on these fields, eliminating one degree of freedom. We will discuss this first for massive particles, one might think for example of the neutral vector boson of weak interactions, or of massive photons, having three degrees of freedom.

Consider the expression for a real vector field,

$$A_\mu(x) = \sum_{\vec{k}} \frac{1}{\sqrt{2Vk_0}} \sum_{i=1}^{3} \left\{ e_\mu^i(k) a^i(\vec{k}) e^{ikx} + \bar{e}_\mu^i(k) a^{\dagger i}(k) e^{-ikx} \right\}$$

with as usual $k_0 = \sqrt{\vec{k}^2 + \kappa^2}$. Further, $\bar{e}_\mu$ equals the complex conjugate of e_μ except that the i of the fourth component is not conjugated, i.e., $\bar{e}_\mu = (e_\mu)^*$ for $\mu = 1, 2, 3$ and $\bar{e}_4 = -(e_4)^*$. This expression is equally valid for massive or massless vector particles; for massless vector particles such as the photon simply put the mass $\kappa = 0$. The polarization vectors e_μ and $\bar{e}_\mu$ (not e_μ^*) transform as vectors under Lorentz transformations. They must be normalized, that is $\bar{e}_\mu^i e_\mu^i = 1$ for $i = 1$, 2 or 3. As usual that corresponds to one particle in the universe.

Restricting the number of degrees of freedom means restricting the number of different polarization vectors $e_\mu^i(k)$. The essential point is to restrict them in a way consistent with Lorentz invariance. For example, the restriction to polarization vectors that have the first component zero is not consistent with Lorentz invariance, because a suitably chosen rotation turns a vector with zero first component into a vector with non-zero first component. A good way is for example to require that the polarization vectors are orthogonal to the four-momentum k, thus $k_\mu e_\mu^i(k) = 0$. That property remains true under a Lorentz transformation, because dot-products are Lorentz invariant. This limits the number of possible polarization vectors to three. That is the normal situation for vector particles: three degrees of freedom. The eliminated possibility, $(0, 0, 0, i)$ in the k rest frame, is actually quite bad from a physics point of view. The relativistic i gets in the way, and it is not possible to normalize this polarization vector. This leads to negative probabilities. Take note: we would not be able to formulate a theory with four degrees of freedom. One of them would be unnormalizable, and we could not talk of the probability of observing that one. For the massless case we will see that two degrees of freedom are similarly unphysical.

Exercise 7.1 Try to normalize ($e_\mu \bar{e}_\mu = 1$) this fourth polarization vector.

7.2 Subsidiary Condition Massless Vector Fields

For a massless photon the energy equals the absolute value of the momentum, $E = |\vec{k}|$, as is clear from the energy–momentum relation above. Thus the four-momentum of the photon with momentum $\vec{k}$ moving along the third axis is $(0, 0, |\vec{k}|, i|\vec{k}|)$. The dot-product of this vector with itself is zero: $k^2 = \vec{k}^2 - \vec{k}^2 = 0$. Thus the condition that the polarization vector be orthogonal to the momentum k would have as a solution also k itself! This is not acceptable, because the dot-product of such a polarization vector with itself would be zero as well, and polarization vectors must be normalized to 1 (that normalization translates directly into probability for finding one such photon). That is now a new problem that presents itself: restricting the number of degrees of freedom by means of the subsidiary condition $e_\mu k_\mu = 0$ allows one solution that is not physically acceptable. Let us consider this in detail.

Consider a massless photon, of momentum k, moving along the third axis. The condition $e_\mu k_\mu = 0$ allows three solutions, for which we may take $e^1 = (1, 0, 0, 0)$, $e^2 = (0, 1, 0, 0)$ and $e^3 = (0, 0, 1, i)$. As stated before, the last one cannot be normalized as it has length zero. Physical photons correspond to the first two polarizations, and we must declare the third one to be unphysical. The problem is now that even if we start with some physical photon, a perfectly reasonable Lorentz transformation may lead to an admixture involving this unphysical one! Let us make this explicit. Consider a sequence of two Lorentz transformations, the first one corresponding to moving with speed v along the first axis,

$$
L_1 = \begin{bmatrix} \frac{1}{\beta} & 0 & 0 & \frac{iv}{\beta} \\ 0 & 1 & 0 & 0 \\ 0 & 0 & 1 & 0 \\ \frac{-iv}{\beta} & 0 & 0 & \frac{1}{\beta} \end{bmatrix}
$$

with $\beta = \sqrt{1 - v^2}$. This transforms the four-vector k into the vector $|\vec{k}| \times (-v/\beta, 0, 1, i/\beta)$. It has obtained a non-zero first component. We subsequently perform a rotation around the introduce a fictitious particle, non-interacting, and construct the field Λ from that. Gauge invariance implies then that physics remains unchanged if we add an arbitrary amount of this field to the electromagnetic field. Here is the first suggestion of a ghost field, typical for gauge theories.

Thus, given a photon of momentum $k = (0, 0, |\vec{k}|, i|\vec{k}|)$ there are two physical polarization vectors, for which we could take for example $(1, 0, 0, 0)$ and $(0, 1, 0, 0)$. This then presumably corresponds to the massless photon case, with its two degrees of freedom. The third degree of freedom (polarization vector proportional to k) must decouple.

7.3 Photon Helicities

The choice $(1, i, 0, 0)/\sqrt{2}$ and $(1, -i, 0, 0)/\sqrt{2}$ for the polarization vectors of the massless vector field is also possible, and corresponds directly to photons with spin along, respectively opposite to, the direction of motion of the photon. Note the normalization, $e_\mu \bar{e}_\mu = 1$. These polarization vectors have a surprising property: under a Lorentz transformation that leaves the momentum k the same these vectors do not transform into each other. They transform into themselves with possibly an additional piece along k, as shown above. They are independent degrees of freedom, and one could actually formulate a theory containing one kind but not the other. More about that later. Here we wish to concentrate on the property that they are eigenvectors for a rotation around the third axis (the direction of motion of the photon), since that provides us with the value of the spin along the third axis. For a rotation over an angle ϕ the first vector has an eigenvalue $\exp(i\phi)$, the other $\exp(-i\phi)$. The factor in front of $i\phi$ is the value of the spin along the third axis. That is thus $+1$ and -1 respectively. The conservation of spin along the third axis in any reaction follows from invariance under rotations around the third axis. Thus for all particles combined all factors $\exp(i\lambda\phi)$ must cancel. It is instructive to see how spinors behave under rotations around the third axis! If the above is true they should obtain factors such as $\exp(i\phi/2)$.

Exercise 7.2 Consider the spinor table given in an earlier chapter. Take the particles at rest (all p_j zero). Check the behaviour of these spinors under a rotation around the third axis.

This is a somewhat complicated exercise. Use an infinitesimal rotation. For a rotation around the third axis over an angle ϕ the Lorentz transformation has all $\alpha^{\mu\nu}$ zero except $\alpha^{12} = \phi$ and $\alpha^{21} = -\phi$. Write down the corresponding spinorial transformation X, and, given that ϕ is infinitesimal, expand the exponential. Use the explicit form of the γ-matrices to obtain σ^{12} explicitly. Apply the resulting matrix to the spinors, and obtain the eigenvalue $(1 + i\lambda\phi)$, with λ the factor corresponding to the value of the spin along the third axis. For the anti-particle spinors the result appears to be the opposite of that expected, but that has to do with the meaning of these spinors (representing outgoing rather than incoming particles). If you want to understand that study the decay of a pseudoscalar particle A into an electron and a positron with the interaction Hamiltonian $A(x)(\bar{\psi}(x)\gamma^5\psi(x))$. Consider the case that the masses of the particles are such that the three-momentum of the outgoing fermions is very small and may be neglected. Then the outgoing fermions must have opposite spin directions, as in that case there can be no orbital angular momentum and the angular momentum of the initial state is zero. Of course, particles having some orbital angular momentum with respect to each other must move with some speed with respect to each other. Positronium (bound state of electron and positron) in the $^1S_0$ state is like a pseudo-scalar (that is why there is a γ^5) and decays like the A. Without the γ^5 the decay would be zero in the limit of small three-momentum, meaning that in that limit a scalar particle cannot decay into two fermions of spin 1/2.

7.4 Propagator and Polarization Vectors of Massive Vector Particles

Treating the photon as a massive particle in the limit of zero mass is not very satisfactory. Of course, we do not know if the photon is truly massless or if it has a very, very small mass. The main issue is then this: is there, theoretically, a difference between the two cases? The answer is: perhaps. We have already seen that there is a difference with respect to the number of degrees of freedom. Physically there is the following fundamental difference between a finite mass photon and a zero mass photon. A finite mass photon is something that will have a speed less than the speed of light (well, that is really a contradiction in terms, but we assume that the reader knows what we mean). By speeding yourself up to sufficiently large speed you could eventually catch up with that photon. In other words, there exists a Lorentz transformation that transforms such a photon to a photon at rest. This is never possible with a massless photon. That is a discrete difference between a massless and massive photon.

There is a more explicit discrete difference. A photon is a particle with spin one, and in its rest frame such a massive particle has three polarization states. By suitable rotations any state can be rotated into any other. A massless photon has only two polarization states, namely spin along or opposite to the direction of motion. Moreover, they are independent, one can not rotate one into the other. To rotate one into the other requires an axis of rotation perpendicular on the direction of motion of the photon; such a rotation however also reverses the direction of motion of the photon and spin along that direction remains spin along that direction. In conclusion, the massive and massless case are vastly different.

The propagator for a massive vector particle is (we ignore contact terms. See Unitarity, section on momenta in propagator):

$$\frac{\delta_{\mu\nu} + k_\mu k_\nu / M^2}{k^2 + M^2 - i\epsilon}$$

The expression in the numerator relates directly to the sum over spins of the product of polarization vectors:

$$\sum_i e_\mu^i(k)\bar{e}_\nu^i(k) = \delta_{\mu\nu} + \frac{k_\mu k_\nu}{M^2}.$$

This relation is required by unitarity, something to be discussed in detail later.

We have already mentioned the fact that this propagator, behaving as a constant in the limit of large k, is bad from the point of view of renormalizability. Here we want to concentrate on the behaviour for small mass, $\kappa \to 0$. That the limit of zero mass of a massive photon is not trivial is something that is very clear from the form of this massive photon propagator. For κ going to zero we have clearly a problem. The issue is evaded by coupling the photon to the electron in such a way that the $k_\mu k_\nu$ term gives zero. This may be formulated in the following way: make the photon interaction with electrons such that the third state of polarization (zero spin along the direction of motion) decouples from the photon. In that way the limit zero mass becomes the zero mass case. The question is if indeed this decoupling can be achieved. The reason that this is questionable is simply Lorentz invariance. For a massive particle spin states may be transformed into each other: first go to the rest frame, then rotate. So, decoupling one spin state automatically implies decoupling them all. At best one can decouple something that becomes very near to this third polarization state in the case of zero mass.

For a vector particle of mass M at rest there are three spin states, described by the four-vectors e_μ^j, $j = 1, 2, 3$, with:

$$e_\mu^1 = (1,0,0,0), \quad e_\mu^2 = (0,1,0,0), \quad e_\mu^3 = (0,0,1,0).$$

Let us make a Lorentz transormation along the third axis, such that after that the particle has momentum $\vec{k}$ in that direction. The first and second polarization vectors, being along the first and second axes, remain the same. The third becomes:

$$e_\mu^3 = (0,0,\frac{E}{M},\frac{i|\vec{k}|}{M}), \quad E = \sqrt{\vec{k}^2 + M^2}.$$

Exercise 7.3 Verify this.

In the limit of very large momentum $\vec{k}$ with respect to the mass M this polarization vector becomes very nearly equal to the vector $k_\mu/M = (0,0,|\vec{k}|,iE)/M$:

$$e_\mu^3 - \frac{k_\mu}{M} = \frac{E - |\vec{k}|}{M} v_\mu, \quad v_\mu = (0,0,1,-i).$$

Since for large $|\vec{k}|$ we may write

$$E = \sqrt{\vec{k}^2 + M^2} \approx |\vec{k}| + \frac{M^2}{2|\vec{k}|} + \cdots$$

we see that in the limit of small mass the difference between this polarization vector and k_μ/M becomes zero. If we can choose the coupling of the photon to the electron such that anything proportional to k_μ (the four-momentum of the photon) decouples then that may be good enough. The four-vector k_μ is different from the polarization vectors in the sense that no Lorentz transformation transforms it into any of the polarization vectors. This may be seen as follows. Go to the rest system. The polarization vectors have obviously length 1, and are equally obviously orthogonal to the four-vector $k_\mu = (0, 0, 0, iM)$. Since the dot-product $(e\,k)$ is Lorentz invariant it remains zero under a Lorentz transformation, and k_μ remains orthogonal to the polarization vectors.

The curious thing here is that the third polarization vector becomes arbitrarily close to k_μ under a suitable Lorentz transformation and yet remains orthogonal. That is typical for the strange things that happen in Minkowski space. Note that the individual components of this third polarization vector become very large for high energy (or small mass). Yet its length, i.e., the dot-product with itself remains one.

The vector v_μ is a strange thing as well: it has length zero, i.e., $(v\,v) = 0$.

It is clear that delicate things are going on. This third polarization vector is a very dangerous thing, its components may become very large. One wrong move and you have a problem in the zero mass limit. Indeed, the rule is that it will not decouple. Only in the case of quantum electrodynamics can decoupling be achieved.

Similar things happen for spin-two particles such as the graviton. The limit of zero mass of a massive graviton is not equal to the zero mass case, and from the experimental observations one may actually deduce that the massless graviton is what nature uses.

The first priority is now to derive some expression for the propagator of a massless photon.

7.5 Photon Propagator

The trouble with a propagator is that the associated momentum squared k^2 is generally non-zero, also for zero mass photons. In that sense the propagating photon is a massive particle. A massive photon has three polarization states, and we are clearly heading for a clash with Lorentz invariance. This is because it is Lorentz invariance that implies three polarizations for a massive particle: go to the rest frame, and then by rotation a polarization vector can be transformed into the other two.

Let us see this explicitly. Consider a photon moving along the third axis. The two possible polarization vectors are $(1,0,0,0)$ and $(0,1,0,0)$. To construct a propagator we need the sum over polarizations, and we find:

$$\sum_j e_\mu^j e_\nu^j = \begin{pmatrix} 1 & 0 & 0 & 0 \\ 0 & 1 & 0 & 0 \\ 0 & 0 & 0 & 0 \\ 0 & 0 & 0 & 0 \end{pmatrix}.$$

This may actually be formulated in terms of the vector k_μ if we first introduce the new four-vector $\tilde{k}_\mu$ where $\tilde{k}$ is obtained from k by giving its spatial components (the first three) a minus sign. One has:

$$\sum_j e_\mu^j e_\nu^j = \delta_{\mu\nu} - \frac{\tilde{k}_\mu k_\nu + k_\mu \tilde{k}_\nu}{(k\tilde{k})}$$

This is easily verified; note that $k = (0,0,|\vec{k}|, i|\vec{k}|)$ and the dot-product $(k\tilde{k}) = -2\vec{k}^2$. All this holds only on the mass shell, where k has the form shown. But that is generally all one needs for unitarity, as will be discussed later.

This formal expression for the sum over polarizations remains correct even if the direction of $\vec{k}$ is not along the third axis, because the definition of $\tilde{k}$ is invariant under rotations. By this we mean that under a rotation a vector in the opposite direction of $\vec{k}$ becomes a vector in the opposite direction of the transformed $\vec{k}$. Unfortunately, the expression for the sum over polarizations is not Lorentz invariant, because the definition of $\tilde{k}$ is not Lorentz invariant. Take for example a particle moving with momentum $\vec{k}$ along the positive third axis. The vector $\tilde{k}$ is the momentum of a particle moving in the opposite direction, towards the negative third axis. Now make a Lorentz transformation along the third axis: the momentum of the particle moving upwards will decrease, that of the other one increase.

Formally this can also be put as follows. Since $(k\tilde{k}) = 2|\vec{k}|^2$, and since a Lorentz transformation changes the magnitude of $\vec{k}$ we see that this dot-product is not invariant, i.e., $\tilde{k}$ does not transform as a Lorentz vector.

Thus, for a given coordinate system we can use the expression (ignoring contact terms):

$$\frac{\delta_{\mu\nu} + \dfrac{\tilde{k}_\mu k_\nu + k_\mu \tilde{k}_\nu}{2\vec{k}^2}}{k^2 - i\epsilon}$$

as a propagator for the photon. But this is not Lorentz invariant. The Lorentz invariance of the S-matrix computed with this expression for the photon propagator becomes questionable. In other words, if one computes things in another coordinate system one may obtain a different result.

But there is hope. Again we see that the problems may disappear if the k_μ terms in the propagator may be neglected. So, both in the massive and massless case the crucial question seems to be the same: one can ignore these terms containing k_μ (or k_ν). There is a subtle difference: in the massless case the dangerous terms contain a single k (and a $\tilde{k}$), in the massive case there are two k's. In Yang–Mills theories that difference is important. In fact, the massless case was understood before the massive case was understood.

It is important to understand that the troubles are very different. In the massless case the dangerous terms give rise to a violation of Lorentz invariance. In the massive case there is no problem with gauge invariance, but the limit of zero mass cannot be taken. Moreover, for the massive case, the removal of the $k_\mu k_\nu$ terms is essential with respect to renormalizability.

In general one needs special properties for the coupling of massless vector particles to other particles such as electrons or neutrinos, or else the theory is not Lorentz invariant. This special property is gauge invariance. If gauge invariance holds then it is possible to derive relations for diagrams that can be used to eliminate the dangerous propagator terms.

So keep this in mind. A theory with massless vector particles, such as quantum electrodynamics or quantum chromodynamics, must have a gauge invariance else the theory is not Lorentz invariant. Theories with massive vector particles need gauge invariance or else the theory is not renormalizable.

In field theory you can usually shift a difficulty around, such as repairing Lorentz invariance but getting some other disease in exchange. For example, leaving out the difficult part of the propagator restores Lorentz invariance, but now the S-matrix is no more unitary, i.e. conservation of probability is lost.

The required relations for diagrams are called Ward identities, for Yang–Mills theories one speaks also of generalized Ward identities or Slavnov–Taylor identities. Understanding this precisely requires a considerable amount of work.

7.6 Left Handed Photons

It is actually possible to formulate a theory that contains photons of one helicity (say, left handed photons, spin opposite to the direction of motion), but not the other. Nothing of that kind has been observed thus it is at this point merely a curiosity. The resulting theory is non-renormalizable.

Take as usual the photon moving along the positive third axis. The polarization vector corresponding to a photon with spin along the negative third axis is given by $e_\mu^L = (1, -i, 0, 0)/\sqrt{2}$. The sum over polarizations generates the matrix to be used in the propagator:

$$e_\mu^L \bar{e}_\nu^L = \tfrac{1}{2} \begin{pmatrix} 1 & i & 0 & 0 \\ -i & 1 & 0 & 0 \\ 0 & 0 & 0 & 0 \\ 0 & 0 & 0 & 0 \end{pmatrix}.$$

As it happens this matrix can be generated using the same vector $\tilde{k}$ introduced before (obtained from k by giving a $-$ sign to the spatial components). For a propagator one needs the matrix for the particle going one way and also for the particle going the other way, which normally is the same, but not here:

$$\frac{1}{k^2 - i\epsilon} \tfrac{1}{2} \left(\delta_{\mu\nu} + \frac{k_\mu \tilde{k}_\nu + \tilde{k}_\mu k_\nu}{2\vec{k}^2} + (\theta(k_0) - \theta(-k_0)) \frac{\epsilon_{\mu\nu\alpha\beta} k_\alpha \tilde{k}_\beta}{2\vec{k}^2} \right).$$

There is another tricky point concerning the use of this expression off mass shell. In general one must use there for k_0 either k_0 itself or else $+\sqrt{\vec{k}^2 + m^2}$, where m is the mass of the particle, here 0. In the case at hand one must use the latter for the k_0 contracted with the ϵ, and that is really the meaning of the θ functions. Careful study of the time-ordered product, $\Delta_F(x) = \theta(x_0)\Delta^+ + \theta(-x_0)\Delta^-$, is necessary to understand these unpleasant matters.

The photon with the opposite helicity would have given a similar expression with the sign of ϵ changed. The tensor ϵ is of course the usual completely antisymmetric tensor.

If we are to have a Lorentz invariant theory with a propagator numerator of this form then the terms involving $\tilde{k}$ must somehow disappear. Gauge invariance will take care of the terms containing k_μ or k_ν, but the ϵ term will remain. There is only one solution: use for the propagator both helicities, but use interactions that produces only left handed photons. This is actually possible by using in the interactions exclusively the combination $F_{\mu\nu} + \tilde{F}_{\mu\nu}$ where $F_{\mu\nu} = \partial_\mu A_\nu - \partial_\nu A_\mu$, and $\tilde{F}_{\mu\nu} = \epsilon_{\mu\nu\alpha\beta}F_{\alpha\beta}$. We will show how that works on the example of the Yndurain interaction (photon interacting with muon and electron):

$$\mathcal{H}_Y = -\frac{g}{2}\left(F_{\mu\nu} + \tilde{F}_{\mu\nu}\right)(\bar{\psi}_\mu \sigma^{\mu\nu}\psi_e) = gA_\mu\partial_\nu\left(\bar{\psi}_\mu\sigma^{\mu\nu}(1-\gamma^5)\psi_e\right).$$

One must add the hermitian conjugate interaction, but we will concentrate on this vertex. In the above σ is the usual antisymmetric combination of two γ-matrices. The interaction is gauge invariant, and it leads to the vertex factor

$$-\frac{ig}{4}k_\nu\left(\gamma^\mu\gamma^\nu - \gamma^\nu\gamma^\mu\right)\left(1-\gamma^5\right) \equiv -\frac{ig}{4}X_\mu.$$

We will show that the interaction multiplied with the propagator expression for the other photon (that is thus the above shown expression but with opposite sign for the ϵ tensor) gives zero. Then we can use for the photon propagator the sum of both, which gives the usual photon propagator, just $\delta_{\mu\nu}$ in the numerator.

Consider now the vertex factor X_μ for a photon of momentum k, on mass shell, with $\vec{k}$ along the third axis and positive k_0 incoming. For unitarity we need only on mass-shell behaviour. Note, however, nothing is assumed about the fermions, they may well be off mass shell. Using equalities such as $\gamma^1\gamma^3\gamma^5 = \gamma^2\gamma^4$ one finds:

$$X_1 = 2|\vec{k}|\left(\gamma^1\gamma^3 - \gamma^2\gamma^4 + i\gamma^1\gamma^4 + i\gamma^2\gamma^3\right);$$
$$X_2 = 2|\vec{k}|\left(\gamma^2\gamma^3 - \gamma^1\gamma^4 - i\gamma^2\gamma^4 - i\gamma^1\gamma^3\right);$$

$$X_3 = 2i|\vec{k}| \left(\gamma^3\gamma^4 + \gamma^1\gamma^2 \right) ;$$

$$X_4 = -2i|\vec{k}| \left(\gamma^3\gamma^4 + \gamma^1\gamma^2 \right) .$$

The components (X_1, X_2) may be taken together and we call that the vector V_μ, having zero third and fourth components. Similarly (X_3, X_4) constitutes the vector W_μ. One has:

$$V_\mu = 2|\vec{k}| \left(\gamma^1\gamma^3 + \gamma^2\gamma^4 - i\gamma^1\gamma^4 + i\gamma^2\gamma^3 \right) \ell_\mu^- ;$$

$$W_\mu = 2i \left(\gamma^3\gamma^4 + \gamma^1\gamma^2 \right) k_\mu$$

with $\ell_\mu^- = (1, -i, 0, 0)$. The W_μ term gives zero when multiplied into the propagator or a polarization vector, the quantity ℓ_μ^- gives zero when multiplied with the right handed propagator or right handed polarization vector, $(1, -i, 0, 0)$. Thus only a left handed photon is produced.

8
Unitarity

8.1 U-matrix

For all practical purposes the Feynman rules represent the true content of a theory. Knowing them we can compute various quantities and compare them with the data. There were a number of unsatisfactory things in the way we derived them, the principal being the U-matrix. There exist even formal proofs that this U-matrix does not exist!

In massless quantum electrodynamics it becomes much more difficult to follow the path of equations of motion, U-matrix, S-matrix. The problem is that the equation of motion for the electromagnetic field allows too many solutions. That equation is:

$$\partial_\mu F_{\mu\nu} = -j_\nu$$

with

$$F_{\mu\nu} = \partial_\mu A_\nu - \partial_\nu A_\mu.$$

The equation makes no sense unless $\partial_\nu j_\nu = 0$, as may be seen by applying ∂_ν to that equation. Furthermore the equation is gauge invariant, which means that, given that A_μ is a solution, then also $A_\mu - \partial_\mu \Lambda(x)$ is a solution with an arbitrary (differentiable) function Λ. This is obvious if we realize that $F_{\mu\nu}$ is invariant for such a replacement. It follows that the solution of the equations of motion is not unique. To restrict the solutions one usually formulates a subsidiary condition, such as for example the Lorentz condition $\partial_\mu A_\mu = 0$, and one limits onseself to those solutions that satisfy this subsidiary condition. If we are to follow this same procedure then one must somehow formulate a subsidiary condition for the U-matrix. This turns out to be very difficult and gives rise to really ugly things. Moreover, while in classical theory one has a wide choice of subsidiary conditions, there is nothing like it in the quantum case for the U-matrix.

In addition, the treatment of the previous chapter shows that there are further serious troubles, in particular the form of the photon propagator is unclear. It is not something that can be derived easily using the operator formalism, which is obvious to us, since we know that there is possibly a conflict with Lorentz invariance unless the interactions satisfy special properties.

The solution is simply to do away with this whole U-matrix thing and equations of motion. Look at it this way. The equations of motion were assumptions, and whether they are correct or not follows in the end, by comparison of the theory with the data. The relevant thing for the S-matrix is the interaction Hamiltonian, which determines the vertex structure. Why all these manipulations of operators, commutation rules and what not if in the end the only relevant thing is the interaction Hamiltonian? Why not just define an interaction Hamiltonian, and thus vertices, and take it from there?

Well, there is one important reason. The U-matrix, and perforce the S-matrix, is unitary by construction (if the interaction Hamiltonian is hermitian). If we start directly with the Feynman rules as defined by an (assumed) interaction Hamiltonian then we do not know if the resulting S-matrix is unitary. This is truly important, because unitarity implies conservation of probability, and probability is the link between the formalism and observed data. If we have no unitarity, then we have nothing that can be interpreted as probability and the link to observed processes disappears. In any case, as we have argued before, things like unitarity, Lorentz invariance, locality etc. are in some sense interchangeable, and these properties must be seen as a framework within which the theory must be formulated.

If we abandon the U-matrix formalism then we must have some other means of establishing unitarity of the S-matrix. Actually, also some form of causality (a notorious difficult thing in quantum mechanics) as well as Lorentz invariance must be established, but that is usually not too hard.

It is thus necessary to obtain a good insight in the relationship between Feynman rules and unitarity of the S-matrix. The crucial tool here is the largest time equation.

8.2 Largest Time Equation

The largest time equation holds for any individual diagram constructed with propagators that satisfy certain simple properties. For the moment we will assume the simplest possible propagators, that is those of scalar particles:

$$\Delta_F(z) = \frac{1}{(2\pi)^4 i} \int d_4 k \, \frac{e^{ikz}}{k^2 + M^2 - i\epsilon}.$$

The mass appearing in this expression is of no importance. Every propagator may have a different mass, or all the same, never mind. Crucial is the $-i\epsilon$. Also, we will assume for the moment that the expressions for the vertices are just some coupling constants, apart from the usual factor $i(2\pi)^4$.

A propagator as the above follows from the expression:

$$\Delta_F(z) = \theta(z_0)\Delta^+(z) + \theta(-z_0)\Delta^-(z),$$

with

$$\Delta^{\pm}(z) = \frac{1}{(2\pi)^3} \int d_4 k \, e^{ikz} \theta(\pm k_0)\delta(k^2 + M^2).$$

As is obvious from this expression Δ^- is the complex conjugate of Δ^+ and vice versa. For the complex conjugate of the Feynman propagator we thus have:

$$\Delta_F^*(z) = \theta(z_0)\Delta^-(z) + \theta(-z_0)\Delta^+(z).$$

The expression for Δ_F tells us that for $z_0 > 0$ it is equal to Δ^+ and for $z_0 < 0$ it is equal to Δ^-. The largest time equation is a direct generalization of this for an expression containing many propagators.

Consider now some diagram. There are n vertices at the space-time points $x_1 \ldots x_n$. The points are connected by propagators. The argument of the propagator connecting points x_i and x_j is $x_j - x_i$.

Let now the time of one of the x, say x_m, be the largest time. Thus $(x_m)_0 > (x_i)_0$ for all $i \neq m$. Then for all propagators containing as argument $x_m - x_i$ (or $x_i - x_m$) we may replace Δ_F by Δ^+ (or Δ^-). Doing that gives the largest time equation for the simplest case. It will be very helpful to put this in terms of diagrams. For this purpose we need to define some extra rules, because now internal lines may be Δ^+ or Δ^-.

Assume now that certain vertices have a circle drawn around

them (in the above case that would be the vertex at point x_m). We now define the following rules:

- Anticipating complex conjugation we will give a circled vertex a minus sign.
- To a line connecting two uncircled vertices corresponds a Feynman propagator Δ_F.
- To a line connecting an uncircled to a circled vertex corresponds a Δ^+.
- To a line connecting a circled to an uncircled vertex corresponds a Δ^-.
- To a line connecting two circled vertices corresponds the complex conjugate of the Feynman propagator, Δ_F^*.

The general form of largest time equation is now as follows. Let the time $(x_m)_0$ be the largest. Consider now any diagram, with any number of vertices circled. Then the diagram obtained by either removing or adding a circle around the point x_m equals minus the original diagram. The minus arises because of the convention that a circled vertex obtains a minus sign as compared to the uncircled one.

Some examples are in order here. The very simplest case is a diagram containing only one line, connecting two points x_1 and x_2. There is only one propagator, $\Delta_F(x_2-x_1)$. Let $(x_2)_0 > (x_1)_0$. Then the following two equations hold:

$$\Delta_F(x_2 - x_1) = - \Delta^+(x_2 - x_1)\,(-1)$$

$$(-1)\,\Delta^-(x_2 - x_1) = - (-1)\,\Delta_F^*(x_2 - x_1)\,(-1)$$

The reader will have no trouble verifying the correctness of these equations. Remember the minus sign of the circled vertex.

A self-energy diagram corresponds to the product of two propagators. The example below is really the same as the previous, except we have the propagators squared. Assume that the time of the point associated with the rightmost vertex is the largest. Then the equations depicted in the figure hold.

second axis over an angle ϕ such that the first component is rotated away. This is achieved with an angle ϕ such that $\sin \phi = v$ and $\cos \phi = \beta$:

$$L_2 = \begin{bmatrix} \beta & 0 & v & 0 \\ 0 & 1 & 0 & 0 \\ -v & 0 & \beta & 0 \\ 0 & 0 & 0 & 1 \end{bmatrix}$$

This changes the vector k to $|\vec{k}| \times (0,0,1/\beta, i/\beta)$, which differs from the original k only by a factor $1/\beta$.

However, the first polarization vector, $(1,0,0,0)$, becomes first $(1/\beta, 0, 0, -iv/\beta)$ and then $(1, 0, -v/\beta, -iv/\beta)$. This is equal to the original polarization vector plus a small admixture of the undesirable third one:

$$e_\mu^1 \to e_\mu^1 - \frac{v}{\beta} e_\mu^3.$$

The length of this polarization vector remains equal to one.

Thus it is not possible to exclude the "bad" third polarization vector by any Lorentz invariant procedure. There is only one option left: we must make the interactions such that this third photon decouples. This is achieved by gauge invariance. In classical electromagnetism the statement of gauge invariance is that some electromagnetic field A_μ describes the same physics as the field $A_\mu + \partial_\mu \Lambda$, with Λ an arbitrary differentiable real function of space-time. Now ∂_μ leads, for the Fourier transform of Λ, to a factor ik_μ, that is like a polarization vector proportional to k. Thus gauge invariance implies the irrelevance of this third polarization vector, which is proportional to k.

In conclusion, for massless vector fields the subsidiary condition is not sufficient to exclude unphysical effects. In addition the interactions must be constrained such that one more degree of freedom is eliminated from the real world. In classical theory that constraint is gauge invariance.

In quantum theory this is more complicated. The field A_μ is an operator, a matrix in Hilbert space. What does gauge invariance mean? What must be taken for the gauge function Λ? Another field? But for what kind of particle? Already now we can see that:

Here a complicated case, with the time associated with the top vertex the largest. We moved both diagrams to the left of the equal sign.

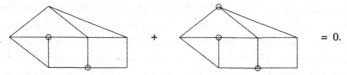

This then is the largest time equation. Its derivation uses only the properties of the θ functions as occurring in the expressions for Δ_F and Δ_F^*. The $-i\epsilon$ in the Fourier transform of the Feynman propagator reflects these θ functions, the origin of this $-i\epsilon$ is in the expression for the Fourier transform of a θ function:

$$\theta(z) = \frac{1}{2\pi i} \int d\tau \frac{e^{i\tau z}}{\tau - i\epsilon}.$$

For scalar particles the propagator is symmetrical in its argument, $\Delta_F(z) = \Delta_F(-z)$. In other words, for a propagator connecting two points x_1 and x_2 one may write either $\Delta_F(x_1 - x_2)$ or $\Delta_F(x_2 - x_1)$. This is fine, as long as this is done consistently also in the associated $\Delta^\pm$ and Δ_F^*.

Exercise 8.1 Verify this for the first example. Note that $\Delta^+(-z) = \Delta^-(z)$.

Now the Fourier transforms of the $\Delta^\pm$ functions also contain θ functions. It is possible to exploit these as well, as we will show now.

8.3 Cutting Equations

As a first step we derive an equation useful if we do not know which time is the largest. Consider any diagram without circles. Now write down the sum of all diagrams obtained by circling vertices in all possible ways. The figure shows the example of a simple self-energy diagram.

The result of this sum is zero, as diagrams cancel two by two, depending on which time is the largest. If for example x_m has the largest time, then any diagram with the vertex x_m not circled cancels against the same diagram with the vertex x_m circled. For example, if the time associated with the right vertex in the above self-energy example is the largest then the first two diagrams cancel against each other, and also the last two cancel. If the left vertex point has the largest time then the first and third diagrams cancel, as well as the second and fourth.

Let us put this slightly more formally. To a diagram with n vertices $x_1, \ldots, x_n$ corresponds some function $F(x_1, \ldots, x_n)$. The same diagram with certain vertices circled can be represented by the function F with the corresponding x underlined: $F(x_1, \underline{x}_2, x_3, \ldots, x_n)$ refers to the same diagram but now with x_2 circled. The above mentioned equation can now be written as:

$$\sum_{\text{underlinings}} F(x_1, \ldots, x_n) = 0.$$

Taking the case of no underlinings and all underlined out from the sum this may be written also:

$$F(x_1, \ldots, x_n) + F^*(x_1, \ldots, x_n) = - \sum_{\text{underlinings } 0, all} F(x_1, \ldots, x_n)$$

where we used that the term with all underlined is the complex conjugate of the term with none underlined. The $0, all$ implies that no underlinings and all underlinings are no more part of the sum.

At this point we move on to considering the S-matrix. That means that we attach to certain vertices external lines, in- or outgoing, and furthermore we integrate all x over all space-time. The equation derived remains true also for the integrated case. Now the diagrams contain the Fourier transforms of the various Δ-functions, and there is conservation of energy–momentum at every vertex. Incoming lines imply positive energy flowing into the diagram, outgoing lines outgoing energy.

What is the effect of the θ-functions in the $\Delta^{\pm}$-functions? These functions occur when a line connects a circled and an uncircled vertex. The effect is very simple: the θ functions are zero unless positive energy flows from the uncircled to the circled vertex.

Exercise 8.2 Verify this for the simplest case, a diagram with just one line between two vertices (of which one is circled) that contain otherwise only external lines.

No restriction exists for the energy flowing through a line connecting two uncircled (a Δ_F-function) or two circled (a Δ_F^*-function) vertices. In summary:

- Positive energy moving through a line connecting an uncircled to a circled vertex can only move towards the circled vertex.
- Energy through lines connecting circled or uncircled vertices may go either way.

Let there now be a circled point in some diagram to which no external line is connected, and where furthermore every attached line connects on the other end to an uncircled vertex, see the figure.

In accordance with the rules concerning energy flow the energy can only flow towards this point. That is not possible in view of energy conservation in that point, and therefore a diagram containing such a configuration is zero.

Obviously the same holds if the point is uncircled, but all points connected to are circled.

This may be generalized from single points to regions. For example, the diagram shown here is zero.

A region may be defined as a connected set of circled (or uncircled) points, connected only to uncircled (or circled) points outside that region. Instead of circled and uncircled vertices we may introduce shadowed lines indicating regions of circled or uncircled points. All points on the shadowed side of such a line supposedly have circles, the others are uncircled. See the figure for an example.

Energy can move across a shadowed line only from the unshadowed towards the shadowed side. The only way that energy can move out from a shadowed region is through an outgoing external line. Thus, a region on the shadowed side must be connected to one or more outgoing lines. Similarly a region on the nonshadowed side must be connected to one or more incoming lines. The shadowed lines cut a diagram into parts connected to in- or outgoing lines. We will by convention draw the shadowed line through the external lines such that the energy flow is from unshadowed to shadowed side.

The statement "sum over all possible circlings" now becomes "sum over all possible cuttings":

$$F + F^* = - \sum_{\text{cuttings}} F$$

The F are functions of the external momenta.

Some examples are in order. First a self-energy diagram.

Note that the cases where no vertex or all vertices are circled must be included as well. We have not specified the energy flow on the external lines; if the left line is ingoing and the right line outgoing then the third diagram is zero. The fourth diagram could equally well have been drawn with the shadowed line going through the right external line and the shadowed side the other way around, that is on the left.

For a triangle diagram with one incoming and two outgoing lines there are five possible cuttings.

The first diagram is the case of no vertex circled, the last with all vertices circled. This last diagram is the complex conjugate of the first.

The equation so obtained holds for any given diagram and is called the cutting  equation. It contains a statement about the real part of any diagram. The entirely shadowed diagram is the complex conjugated of the entirely unshadowed diagram, and the sum of these two, contained in the cutting equation, is twice the real part of the unshadowed (=uncircled) diagram. Twice the real part of a diagram is therefore equal to minus the sum over all cut diagrams. Usually one defines the T-matrix by $S = 1 + iT$, with an extra i. The real part of the diagrams thus contribute to the imaginary part of the T-matrix, and one indeed always refers to the imaginary part of the diagram. Actually, diagrams always have a factor i, thus this usage is not as absurd as it seems at first sight. For example, in the tree diagram in the figure there are two vertices and one propagator, and as each has a factor i there will remain one overall factor i. The self-energy diagram has two vertices and two propagators, but in addition, computing the integral over the closed loop, an i arises from the Wick rotation.

The cutting equation may be used to investigate unitarity of the S-matrix.

8.4 Unitarity and Cutting Equation

Unitarity of the S-matrix means $S^\dagger S = 1$. With $S = 1 + iT$ this means for the T-matrix:

$$i(T - T^\dagger) = -T^\dagger T.$$

We must consider the matrix element of this equation between all possible states, say between states $< b \mid$ and $\mid a >$. The product on the right hand side implies summing over all possible intermediate states:

$$i(< b \mid T \mid a > - < b \mid T^\dagger \mid a >) =$$
$$- \sum_c < b \mid T \mid c >< c \mid T^\dagger \mid a >$$

Understanding the T-matrix in terms of diagrams this equation corresponds closely to the cutting equation. The left hand side is the "imaginary" (after factoring out an i) part of a diagram, the right hand side corresponds to all possible cuttings.

The crucial part is that the $\Delta^{\pm}$-functions represent the phase space for a real particle, on mass-shell and with positive energy. One integrates over all momenta subject to the mass-shell condition and the condition of positive energy.

A simple explicit example is undoubtedly more instructive here than a general argument. Besides, there is really not that much beyond what is already needed at the simplest level. Thus we will consider the σ self-energy diagram in the $\sigma\pi\pi$ model. Moreover, that diagram was already computed before, in the chapter on renormalization.

Let us first get all equations together. The cutting equation here takes the form shown in the figure.

The first diagram was computed before, both with a crude cut-off method and also using dimensional regularization. We are interested only in the "imaginary" part, and only the $\ln A$ part is of relevance. Remember that we must still integrate over the Feynman parameter x. Thus to the left hand side of the cutting equation shown above corresponds the expression (factor $1/2$ for identical particles not included):

$$-2i\pi^2 \, \mathrm{Im} \left[\int_0^1 dx \ln \left(x(1-x)Q^2 + M^2 - i\epsilon \right) \right].$$

By taking out a factor i we now need truly the imaginary part of the logarithm. Actually, many people stick to the convention of always taking out a factor $i\pi^2$, but we prefer to remain very explicit. Remember, Q is the four-momentum passing through the diagram.

Now the x-integral is not very difficult, and we could just compute it. But it is very easy to make mistakes with the imaginary part of the logarithm, and it pays to inspect this expression before evaluating the integral. The logarithm has a cut along the negative real axis, and the $-i\epsilon$ puts the argument just below the cut (if the argument is negative). The imaginary part in that case is simply $-i\pi$, corresponding to an angle of $-180°$ with respect to the positive real axis. Above the cut it would have been $i\pi$. The imaginary part of the whole is the length of the x-region over which the argument is negative times this $-i\pi$, and that is easy to evaluate.

Thus consider the quadratic expression $x(1-x)Q^2 + M^2$. It can be negative provided Q^2 is negative. The maximum of the factor $x(1-x)$ is reached for $x = 1/2$, and then it is $1/4$. The argument can be negative if $Q^2 < -4M^2$, and only then is there an imaginary part. To find the region in x where the argument is negative we must solve the quadratic equation; it is zero for

$$x = \tfrac{1}{2} \pm \tfrac{1}{2}\sqrt{1 + \frac{4M^2}{Q^2}}.$$

It follows that the length of the integration region over which the argument is negative is the difference between these two solutions, which is precisely equal to the square root. We conclude that the imaginary part of the logarithm integrated over x is given by

$$-i\pi \; \theta(-4M^2 - Q^2) \sqrt{1 + \frac{4M^2}{Q^2}}.$$

The θ function makes it zero unless $Q^2 < -4M^2$. Here it is clear how straightforward calculation might give confusion; doing things carelessly one misses the θ function and the answer appears then to have an imaginary part also if $Q^2 > 0$.

The relation $Q^2 < -4M^2$ has a simple physical connection. Go into the Q rest frame (possible if $Q^2 < 0$). Then $Q = (0,0,0,iE)$. The relation in terms of the energy E becomes $E^2 > 4M^2$ or $E > 2M$. In other words, the energy must be larger than twice the π mass, that is there must be sufficient energy to have two real π-particles.

Including all factors, twice the "imaginary" part of the self-energy diagram as depicted in the above cutting equation becomes

$$-2\pi^3 \; \theta(-4M^2 - Q^2) \sqrt{1 + \frac{4M^2}{Q^2}}.$$

Now the right hand side of the cutting equation. Showing all factors including the overall minus sign explicitly the corresponding expression is:

$$-i(2\pi)^4\left[-i(2\pi)^4\right]\frac{1}{(2\pi)^3}\frac{1}{(2\pi)^3}$$

$$\int d_4q\,\theta(q_0)\,\delta(q^2+M^2)\,\theta(Q_0-q_0)\,\delta((Q-q)^2+M^2).$$

Note the minus sign for the vertex in the shadowed region. This integral, which actually has a lot in common with the integral over phase space when computing a two particle decay rate, is best evaluated in the Q rest frame. But it is much more instructive to make this integral resemble the two-body phase space integral. This is achieved as follows. First introduce a new four-integral, but with a δ-function to make it trivial:

$$-(2\pi)^2\int d_4q\int d_4p\,\delta_4(Q-p-q)\theta(q_0)\,\delta(q^2+M^2)\,\theta(p_0)\,\delta(p^2+M^2).$$

Now do the integrals over p_0 and q_0, using up two δ functions. The result of these integrals, done several times earlier, is $1/2p_0$ and $1/2q_0$ with now $p_0=\sqrt{\vec{p}^2+M^2}$ and $q_0=\sqrt{\vec{q}^2+M^2}$. The result is

$$-(2\pi)^2\int\frac{d_3q\,d_3p}{2q_0\,2p_0}\delta_4(Q-p-q).$$

We now have literally the two-body phase space integral including the δ-function for overall conservation of energy–momentum. That including the factors $1/(2\pi)^3$ that come with an integration over final states, and that were here part of the Δ^+-functions.

Exercise 8.3 Show that the integral is zero if $Q^2 > -4M^2$. Hint: use that $Q^2 = (p+q)^2$ due to the δ-function.

At this point we can take over equations derived before. In addition we add a θ-function to express the fact that the integral is zero if $Q^2 > -4M^2$. We simply quote the result:

$$-(2\pi)^2\,\frac{\pi}{2}\,\frac{\theta(-4M^2-Q^2)}{(-Q^2)}\sqrt{(Q^2+2M^2)^2-4M^4}.$$

No factor $1/2$ for identical particles included. This simplifies to

$$-2\pi^3\,\theta(-4M^2-Q^2)\sqrt{1+\frac{4M^2}{Q^2}}.$$

This result coincides with the result found before for the left hand side of the cutting equation.

That the cutting equation holds is good to know, but we must now convince ourselves that the right hand side corresponds correctly to $-T^\dagger T$. That is really the point.

As is clear from the above the calculation of the cut diagram can be written as an integral over two-body phase space. It is therefore the same as what one obtains computing the matrix element of T in absolute value, and then integrating over all states. In this particular case the "in" state is the same as the "out" state, and we have:

$$\sum_c <b \,|\, T^\dagger \,|\, c><c \,|\, T \,|\, b>$$

$$= \sum_c (<c \,|\, T \,|\, b>)^* <c \,|\, T \,|\, b>$$

$$= \sum_c |<c \,|\, T \,|\, b>|^2.$$

The matrix element $<c \,|\, T \,|\, b>$ describes the decay of the $|\, b >$ state into the $|\, c >$ state, here two π particles, and one must integrate over the full two particle phase space.

There is a small point that should be noted here. The original diagram, containing two identical internal lines, should have a combinatorial factor $1/2$. That is precisely what one wants to have for the expression for the cut diagram to be according to the two particle phase space: for two identical particles there is a factor of $1/2$ to account for the fact that the state with the two particles interchanged is the same state. Doing the integral as above one counts double, thus a factor $1/2$ must be added. This then shows also the correctness of this prescription concerning the combinatorial factor.

There is another important fact. We are doing perturbation theory, and in this case we computed the imaginary part of a diagram with two vertices, i.e., of order g^2 if we associate a factor g with each vertex. On the right hand side one has $T^\dagger T$, and one needs here T only to first order in g. In general, to compute the imaginary part of T to order g^m one needs to know T only to order g^{m-1}. Given T up to some order then unitarity alone is sufficient to compute the imaginary part of the next higher order. If we had another tool to compute the real part given the imaginary part then, given the T-matrix in lowest order one can compute the T-matrix iteratively to any order. This program was pursued vigorously in the early sixties; dispersion relations were supposedly the tools to obtain the real part from the imaginary part. This program faltered, as one did not succeed in proving the validity of a sufficiently general set of dispersion relations. For certain simple cases, notably the two-point function (this means diagrams with two external lines, like the one investigated above), dispersion relations hold.

It should be said that the dispersion technique has another application. It might just be that the total decay rate of some particle into all possible states can be measured. That is then the right hand side of the unitarity equation (the $T^\dagger T$ side). That provides us then with the imaginary part of the two-point function. Next, using a dispersion relation the real part may be computed, so that in this way the complete two-point function can be obtained. This has its practical applications with respect to the photon propagator in e^+e^- annihilation as measured at the various big accelerators.

A derivation of the dispersion relation for the two-point function using the largest time equation much like above can be formulated without difficulty. It is usually written in a form called the Källén–Lehmann representation, and will be derived in a later section. Here we will write it down in the form of dispersion relation. The "imaginary" part of the two-point function was computed above, and it is a function of Q^2. Instead of Q^2 we will use the variable $s = -Q^2$. Factoring out the usual factor i we denote the two-point function by $iG(s)$ and the dispersion relation is:

$$G(s) = \frac{1}{\pi} \int_0^\infty ds' \frac{\text{Im } G(s')}{s' - s - i\epsilon}.$$

This expression gives both the real and imaginary part of the two-point function. Note that the integration variable s' runs all the way up to infinity; one must know the imaginary part of G for all s' and that is the catch. No experiment can measure this imaginary part up to infinite energy. Thus the expression is useless unless some estimate can be given for this integral in the limit of high s'. Various methods for attacking this situation can be found in the literature. Good luck.

8.5 Unitarity: General Case

There is no difficulty whatsoever to identify the cutting equation with the unitarity equation for the $\sigma\pi\pi$ theory for arbitrary diagrams. We will now address the complications that arise for other theories, containing particles with spin.

Fermions. The fermion propagator, denoted by S_F, is given by

$$S_F(z) = \frac{1}{(2\pi)^4 i} \int d_4 k \frac{(-i\gamma^\mu k_\mu + m)e^{ikz}}{k^2 + m^2 - i\epsilon}.$$

This propagator differs from the scalar propagator by the factor $-i\gamma^\mu k_\mu + m$, which can also be obtained by differentiation:

$$S_F(z) = (-\gamma^\mu \partial_\mu + m)\Delta_F(z).$$

It follows:

$$S_F(z) = (-\gamma^\mu \partial_\mu + m)\left(\theta(z_0)\Delta^+(z) + \theta(-z_0)\Delta^-(z)\right).$$

We can define functions S^+ and S^-:

$$S^\pm(z) = (-\gamma^\mu \partial_\mu + m)\Delta^\pm(z).$$

Because of the occurrence of z_0 in the θ-functions there is a complication as the derivative of a θ-function is a δ-function. With $\partial_4 = -i\partial_0$ we have:

$$S_F(z) = \theta(z_0)S^+(z) + \theta(-z_0)S^-(z)$$
$$+ i\gamma^4\delta(z_0)\Delta^+(z) - i\gamma^4\delta(z_0)\Delta^-(z).$$

As it happens, for equal times ($z_0 = 0$) Δ^+ and Δ^- are equal, and the two extra terms cancel.

Exercise 8.4 Show that $\Delta^+(z) = \Delta^-(z)$ if $z_0 = 0$.

Complex conjugation becomes roughly hermitian conjugation. We can include fermions in diagrams and derive the largest time equation as before. The only change is to use the S-functions instead of the Δ-functions.

With respect to unitarity, the crucial question is whether the S^+-function gives the same as integration over phase space and spin summation. Now initial and final states contain spinors u and

$\bar{u}$, and we must show that the sum over spins of a product $u\bar{u}$ gives the same as the extra factor in the S^+. For the Fourier transform of S^+ this factor is $(-i\gamma^\mu k_\mu + m)$ and the reader may verify that this is indeed equal to the sum over spins of a spinor combination $u\bar{u}$. This then is the essential point: in the numerator of the Feynman propagator we must have the same thing as obtained by summing over spins of the spinor functions.

Including vector particles offers once more a problem. The Feynman propagator for a massive vector boson is:

$$\Delta_{F\mu\nu}(z) = \frac{1}{(2\pi)^4 i} \int d_4 k \frac{(\delta_{\mu\nu} + k_\mu k_\nu / M^2)\, e^{ikz}}{k^2 + M^2 - i\epsilon}.$$

Also this can be expressed as a derivative:

$$\Delta_{F\mu\nu}(z) = \left(\delta_{\mu\nu} - \partial_\mu \partial_\nu / M^2\right) \Delta_F(z).$$

The situation is more complicated than in the fermion case, we must deal with second derivatives. And while $\Delta^+ = \Delta^-$ for equal times, the same is not true for the derivative of the $\Delta^\pm$. Terms involving $\delta(z_0)$ enter the equations. These terms (called contact terms, as they are non-zero only for equal times) actually do not disturb the derivation of the largest time equation, but proving that is another matter. The reason is that these terms closely relate to the behaviour of the propagator for large k, and study requires regulator methods. Stated differently, one becomes sensitive to the precise details of the θ-functions if the argument is zero. Is it zero, or one, or $1/2$, or the limit of some smooth function?

Apart from this contact term complication, unitarity rests on the identification of the factor $\delta_{\mu\nu} + k_\mu k_\nu / M^2$ with the sum over spins of a product of polarization vectors. Indeed, this sum over polarizations gives that same expression.

There is no point in worrying too much about contact terms for the propagator of vector particles. When dealing with vector particles one needs gauge invariance, and the whole situation changes.

Finally there is the question of vertices. In the largest time equation the vertices are not complex conjugated, the same coupling constants appear in shadowed or unshadowed regions. In order to identify shadowed regions with the matrix $T^\dagger$ (which involves the complex conjugate of the coupling constant), one must either require that the coupling constants are real, or else that next to a vertex with a complex coupling constant there exists the complex conjugate vertex with the complex conjugate coupling constant. In other words, the interaction Hamiltonian (or rather the interaction Lagrangian) must be hermitian.

A theory can thus be non-unitary in the following way. There may be vertices involving complex fields (but real coupling constants), and in $T^\dagger$ one would need the vertices involving the complex conjugate fields. For example, let A and B be complex fields. Then the interaction with a real field C described by the interaction Hamiltonian $gCAB$ leads to a non-unitary S-matrix, unless also the complex conjugated vertex gCA^*B^* is contained in the interaction Hamiltonian. The reason is that $T^\dagger$ looks like using the vertex gCA^*B^*, and that is different from $gCAB$. An example would be the coupling of a neutral vector boson V to a muon and a positron:

$$ig V_\alpha(\bar{e}\gamma^\alpha\mu).$$

We have followed the time honored convention to use the notation μ for the muon field ψ_μ. This allows the decay of this V into a muon and a positron, but not the decay into an electron and an anti-muon, described by the conjugate vertex:

$$ig V_\alpha(\bar{\mu}\gamma^\alpha e).$$

If the first vertex is the only one present that would be the only one occurring in the cutting equations. However, $T^\dagger$ would contain only the conjugate of that vertex, and we cannot identify the cutting equation with unitarity. The theory will be unitary only if both vertices are present.

Note that with the first vertex only there is no simple self-energy diagram for the V: while V may decay into a muon and a positron, the reconversion muon and positron to V is not possible. That needs the second vertex. That shows the problem; while the V can decay, there is no imaginary part in the two-point function. The matrix $T^\dagger$ looks like using the second vertex, which is not available in the cutting equation.

In summary, an S-matrix specified by Feynman rules is unitary if the propagators are of the form described above, closely related to expressions obtained by summing products of spinors or

polarization vectors over spin states. In addition the vertices must
satisfy reality properties.

8.6 Källén–Lehmann Representation, Dispersion Relation

We will now derive the Källén–Lehmann representation for the two-point function. Proving means that we will derive this representation for any self-energy graph, no matter
how complicated.

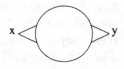

Consider some graph, unspecified except for beginning- and
end-point which we will call x and y respectively. See figure.
If the time y_0 is larger then the time x_0 we can write down again
a relation involving all possible circlings except it is not necessary
to circle x as x_0 is never the largest time. Using the θ-function to
limit ourselves to the case $y_0 > x_0$ we may write:

$$\theta(y_0 - x_0) \sum_{\text{underlinings } x} F(x, y, \ldots) = 0.$$

The notation is obvious: all possible underlinings except x never
underlined. Clearly, we can write a similar equation with y never
underlined and a θ function limiting it to the case $x_0 > y_0$:

$$\theta(x_0 - y_0) \sum_{\text{underlinings } y} F(x, y, \ldots) = 0.$$

The sum includes the case of no underlinings at all. Adding these
two equations we may take the two terms without underlinings
together; these contain the function $F(x, y, \ldots)$ corresponding to
the original diagram without circles. Since $\theta(z) + \theta(-z) = 1$ we
obtain:

$$F(x, y, \ldots) + \theta(y_0 - x_0) \sum_{0,x} F(x, y, \ldots)$$
$$+ \theta(x_0 - y_0) \sum_{0,y} F(x, y, \ldots) = 0.$$

The subscript $0, x$ on the summation symbol implies exclusion
of the term without underlinings and terms with x underlined.
Similarly $0, y$. The word "underlinings" is left understood.

Up to this point the equation is generally valid, not restricted to two-point functions. It is really a dispersion relation in a non-conventional form. See E. Remiddi, Helv. Phys. Acta **54** (1981) 364 for further details. The above equation is his eq. (12), which is the one a few lines below eq. (11). Here we will restrict ourselves to the two-point function.

• To keep track of signs and factors i it is instructive to consider the special case that there is just one line connecting x and y. Including the two factors i for the vertices (the factors $(2\pi)^4$ would arise after integration over x and y) the equation reads:

$$-\Delta_F(y - x) + \theta(y_0 - x_0)\Delta^+(y - x) + \theta(x_0 - y_0)\Delta^-(y - x) = 0.$$

Now integrate over all vertex points except x and y. The θ-functions in the Fourier transform of the $\Delta^\pm$ may be used to work all possible underlinings over to all possible cuttings, just as in the unitarity case. The sum becomes a sum over all possible cuttings. Consider now the function F^+ defined by:

$$F^+(y - x) = \sum_{\text{cuttings } 0,x} f(x, y),$$

where f is F integrated over all vertex points except x and y. The 0 reminds us that the case of no cut is not included. This may be shown diagrammatically, see the figure. F^+ refers to the diagram 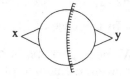 cut in all possible ways, not including x but necessarily always including y since there are no external lines except possibly to be attached to x or y.

We have already indicated that F^+ is a function of $y - x$ only. This because of translation invariance, that is invariance for the replacement $x \to x+a$, $y \to y+a$, with a an arbitrary four-vector. We will actually go to the frame where y is zero.

We are interested in the Fourier transform of F^+. Thus multiply with $\exp(ikx)$ and integrate over all x. That is like attaching an external line with momentum k to the point x. The result is depicted in the figure.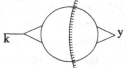

It is evident from the figure that this function is zero unless the energy k_0 is positive, because energy can flow only from unshadowed to shadowed region. However, there is more. Every line cut by the shadowed line is a $\Delta^{\pm}$-function. These functions contain also δ-functions, of the form $\delta(p^2 + M_x^2)$, where p is the momentum flowing through the line, and M_x is the mass of the particle corresponding to that line. In other words, it is as if a real particle on mass shell moves across the cut, which you may remember is precisely the critical point with respect to unitarity. But this means that this diagram, viewed from the incoming k-line to the cut is as if the incoming particle becomes a collection of real on-mass-shell particles with various masses M_x. Kinematically this will go only if the mass of the incoming particle (that is $\sqrt{-k^2}$) is larger than the sum of all masses corresponding to the cut lines. That is $-k^2$ must certainly be positive or else the function is zero. The fact that the function is zero may be indicated explicitly by including θ functions:

$$F^+(-x) = \frac{1}{(2\pi)^4} \int d_4 k \exp(-ikx) \; \theta(k_0) \; \theta(-k^2) \; \sigma(-k^2),$$

with σ some function. A $-$ sign is included in the way we write the argument of σ, anticipating the next step.

• In the case of just one line we have $\sigma = 2\pi \, \delta(k^2 + M^2)$.

We need perhaps to explain some factors here. We computed the Fourier transform of F^+, and let us denote that function by f. Thus:

$$f(k) = \int d_4 z \exp(ikz) F^+(-z).$$

This is the function that is zero unless $k_0 > 0$ and $-k^2 \geq 0$, in fact, it is σ multiplied by the θ-functions. It follows that:

$$F^+(-x) = \frac{1}{(2\pi)^4} \int d_4 k \exp(-ikx) f(k).$$

We know of course perfectly well what the function σ is. Apart from a minus sign it is what appears on the right hand side of the cutting equation used for unitarity. On the left hand side one has the original diagram and its complex conjugate, which we write, factoring out the usual i, as $i(G - G^*)$. This then gives

$$\sigma(-k^2) = 2 \, \text{Im} \, G(k^2).$$

From this we may in particular draw the conclusion that σ is real. If not one diagram but the sum of all possible diagrams is considered then also $\sigma > 0$, see below.

The expression may be rewritten using a new integration over a variable κ and a δ-function:

$$F^+(-x) = \frac{1}{(2\pi)^4} \int_0^\infty d\kappa \, \sigma(\kappa) \int d_4 k \exp(-ikx) \, \theta(k_0) \, \delta(k^2 + \kappa).$$

The restriction to $k^2 < 0$ is now replaced by the restriction that the integration over κ is over positive values only.

The last part of this expression is the Δ^+ function for a particle with mass squared equal to κ. Remembering the factor $1/(2\pi)^3$ in Δ^+ we find:

$$F^+(-x) = \frac{1}{2\pi} \int_0^\infty d\kappa \, \sigma(\kappa) \, \Delta^+(-x, \kappa).$$

Now go back to the beginning. The equation for the whole diagram, integrated over all vertex points except x and y is:

$$-F(y - x) = \theta(y_0 - x_0) \, F^+(y - x) + \theta(x_0 - y_0) \, F^-(y - x).$$

where we introduced the notation F^- for all cut diagrams with x always on the shadowed side and y on the other side. That function may be treated just like F^+; because it is obviously the complex conjugate of F^+ (it is obtained by interchanging shadowed and unshadowed regions) it will involve the complex conjugate of the same function σ as defined for F^+. That function is actually real, as noted before, so the same function σ appears there. The complete expression for $F(y - x)$ is now:

$$-F(z) = \frac{1}{2\pi} \int_0^\infty d\kappa \, \sigma(\kappa) \, \Delta_F(z, \kappa)$$

with $z = y - x$ and furthermore, as usual

$$\Delta_F(z, \kappa) = \theta(z_0) \Delta^+(z, \kappa) + \theta(-z_0) \Delta^-(z, \kappa).$$

This is the Källén–Lehmann representation. Its content may be put in words as follows: the general two-point function can be written as a superposition of Feynman propagators of particles with positive mass squared κ with a weight function $\sigma(\kappa)/2\pi$. One more statement can be made about the function σ: it is positive, at least if one considers the sum of all possible diagrams. The reason is that this σ is simply the absolute value squared of a matrix element. Here we assume the theory to be unitary, so that the right hand side of the cutting equation can be identified with $-T^\dagger T$.

- In the one line case $-F(z) = \Delta_F(z)$ and $\sigma = 2\pi\,\delta(-\kappa + M^2)$.

Finally we will rewrite the above representation as a dispersion relation as discussed in the previous section. This is trivial. Inserting the explicit form of the Feynman propagator we have:

$$-F(z) = \frac{1}{i(2\pi)^4} \int d_4k\ e^{ikz} G(-k^2)$$

where G, the Fourier transform of $-iF$ (writing F as if shows that this is the usual factoring of a factor i), is given by

$$G(-k^2) = \frac{1}{2\pi} \int_0^\infty d\kappa\ \sigma(\kappa) \frac{1}{k^2 + \kappa - i\epsilon}.$$

Now σ is twice the imaginary part of this function G. Using the variables $s = -k^2$ and $s' = \kappa$ the equation for G takes the form of a dispersion relation as given before:

$$G(s) = \frac{1}{\pi} \int_0^\infty \frac{\text{Im } G(s')}{s' - s - i\epsilon}.$$

8.7 Momenta in Propagator

The $\Delta^\pm$-functions may be more complicated then the ones we have used so far, and as specified in the very beginning of this chapter. They may contain more complicated functions of the momentum, for example for fermions they have factors $(-i\gamma k + m)$. For vector bosons there is the factor $\delta_{\mu\nu} + k_\mu k_\nu / m^2$. And for massless photons there are even worse things. But as these factors represent the sum over true physical states they are really crucial. Let us consider here what happens in some more detail.

Thus consider the more general $\Delta^\pm$-functions, involving a function of the four-momentum k. We will explicitly show the dependence on k_0 only, the dependence on $\vec{k}$ is not referred to anywhere and will survive unchanged to the end:

$$\Delta^\pm(z) = \frac{1}{(2\pi)^3} \int d_4k\ e^{ikz} f^\pm(k_0)\theta(\pm k_0)\delta(k^2 + m^2).$$

We introduce the notation $R = +\sqrt{\vec{k}^2 + m^2}$. The δ-function forces $k_0 = \pm R$ and we find:

$$\Delta^{\pm}(z) = \frac{1}{(2\pi)^3} \int d_4 k\, e^{ikz} f^{\pm}(\pm R)\, \theta(\pm k_0)\delta(k^2 + m^2).$$

Now one can follow the lines of the derivation in an earlier chapter up to this:

$$\Delta_F(z) = \frac{1}{(2\pi)^4 i} \int d_4 k\, e^{ikz} \frac{1}{2R}$$

$$\left[\frac{f^+(R)}{-k_0 + R - i\epsilon} + \frac{f^-(-R)}{k_0 + R - i\epsilon} \right]$$

$$= \frac{1}{(2\pi)^4 i} \int d_4 k\, e^{ikz} \frac{1}{2R}$$

$$\left[\frac{R(f^+(R) + f^-(-R)) + k_0(f^+(R) - f^-(-R))}{k^2 + m^2 - i\epsilon} \right].$$

To get the hang of this let us consider a few simple cases. First the case relevant to fermions, $f^{\pm}(k_0) = k_0$, i.e., $f^{\pm}(\pm R) = \pm R$. Then the result is simply k_0. So, in that case terms in $f(k_0)$ linear in k_0 translate to terms k_0 in the propagator numerator. This is also the case for terms linear in k_0 for vector bosons ($\mu = 4$ or $\nu = 4$).

Consider next $f^+(k_0) = f^-(k_0) = k_0^2$, a case relevant to vector bosons for $\mu = \nu = 4$. Then $f^+(R) = f^-(-R) = R^2 = \vec{k}^2 + m^2$, and the resulting Fourier integrand is, using $\vec{k}^2 = k_0^2 + k^2$:

$$\frac{\vec{k}^2 + m^2}{k^2 + m^2} = \frac{k_0^2}{k^2 + m^2} + 1.$$

Thus we get an extra 1, and doing the k integral we have an additional $\delta_4(z)$. This is the contact term mentioned before.

It may be clear that, in general, terms of the form k_0^n translate to a propagator with this same k_0^n in the numerator, but there will be extra contact terms, δ-functions and derivatives of δ-functions.

Finally a nasty case: $f^{\pm}(k_0) = \pm k_0$. Then the result for the Fourier integrand is

$$\frac{\sqrt{\vec{k}^2 + m^2}}{k^2 + m^2},$$

and there is not much one can do about that one. This occurs when dealing with left or right handed photons separately.

It is clear that except for the fermion case there is trouble. In practice a theory must be such that somehow these troublesome terms vanish. For example, in the case of photons one works with the propagator $\delta_{\mu\nu}/(k^2 - i\epsilon)$ without any extra terms. Then, when considering unitarity, there will be $\Delta^{\pm}$-functions that also contain only $\delta_{\mu\nu}$, which is not what is needed, because one must have there what corresponds to the sum over polarizations. If however now it can be shown that the extra terms may be added to these $\Delta^{\pm}$ as they would give zero anyway then all is fine. Thus the extra terms need to vanish only if the photon is on mass-shell.

9
Quantum Electrodynamics: Finally

9.1 Unitarity

A general treatment of unitarity and Ward identities for gauge theories is a complicated affair; here, as a preliminary we will discuss the much simpler case of quantum electrodynamics involving electrons and photons only. We will have then at least for quantum electrodynamics a reasonably complete treatment, showing that the manifestly Lorentz covariant Feynman rules as given for that theory define a Lorentz invariant S-matrix that also conserves probability.

Let us recapitulate the unitarity argument as it applies to this theory. We use the photon propagator

$$\frac{\delta_{\mu\nu}}{k^2 - i\epsilon}.$$

The cutting rules provide us with an equation for the imaginary part of the T-matrix. That equation involves $\Delta^{\pm}$-functions, and associated with the above propagator we will have:

$$\Delta^{\pm}_{\mu\nu}(z) = \frac{1}{(2\pi)^3} \int d_4 k \, e^{ikz} \, \delta_{\mu\nu} \, \theta(\pm k_0) \delta(k^2).$$

The point is now that in order to have unitarity we must have the sum over spins of the possible photon polarization vectors in this expression, and that is not $\delta_{\mu\nu}$ but

$$\delta_{\mu\nu} - \frac{k_\mu \tilde{k}_\nu + \tilde{k}_\mu k_\nu}{(k\tilde{k})},$$

where $\tilde{k}$ is the four-vector obtained from k by space reflection, i.e., giving a minus sign to the first three components. We must therefore show that the extra terms disappear. For this we need Ward identities. Since the $\Delta^{\pm}$-functions connect to diagrams we must show that we get zero if for one or more external photon

207

lines we replace the polarization vector by the associated four-momentum. As we will see, that will not be true diagram by diagram, but only for certain sets of diagrams.

9.2 Ward Identities

As argued above, the last step in demonstrating the correctness of the Feynman rules for quantum electrodynamics is to demonstrate that some identities hold, the Ward identities. These identities then show that the theory is unitary despite the use of a simplified propagator.

The way we will go about this is step by step, explicitly. We start with the simplest case, a diagram containing two external electron lines and one external photon line, see figure. The diagram is actually zero because of kinematics, but that need not bother us here.

Substituting k_μ for the photon polarization vector, the expression associated with this diagram is:

$$ie(2\pi)^4 ik_\mu \left(\bar{u}(q)\gamma^\mu u(p)\right)\delta_4(k+p-q).$$

The procedure should really be very familiar. Write $k_\mu = q_\mu - p_\mu$ and apply the Dirac equation for the spinors:

$$i\gamma p\, u(p) = -m\, u(p), \qquad \bar{u}(q)i\gamma q = -m\,\bar{u}(q),$$

where m is the electron mass. The two terms cancel and we find zero.

So, if the electron lines connected to the photon are external lines then the desired result is very easily found. Just like for PCAC the point is now to show that this works also if the electron lines are propagators. The work is very similar to what we did when computing the anomaly.

Thus consider the case where the line with the replaced polarization vector is attached to an electron line with on both sides a propaga- 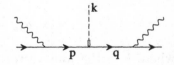 tor. We introduced an explicit no- tation to show that we replaced the polarization vector by the associated four-momentum: a dashed line terminating in a short double piece.

The expression corresponding to this part is:

$$\gamma^\alpha \, \frac{-i\gamma q + m}{q^2 + m^2 - i\epsilon} \, i\gamma^\mu \, k_\mu \, \frac{-i\gamma p + m}{p^2 + m^2 - i\epsilon} \, \gamma^\beta \, .$$

We have not bothered to write all factors such as e or $(2\pi)^4 i$. There is of course conservation of four-momentum: $k = q - p$. We now substitute:

$$i\gamma k = (i\gamma q + m) - (i\gamma p + m) \, .$$

These two terms are actually inverse propagators:

$$\frac{-i\gamma q + m}{q^2 + m^2 - i\epsilon} \, (i\gamma q + m) = 1 \, .$$

In fact, this is historically the way Ward presented this identity:

$$k_\mu V_\mu(p, q) = P^{-1}(q) - P^{-1}(p) \, .$$

The expression becomes:

$$\gamma^\alpha \frac{-i\gamma p + m}{p^2 + m^2 - i\epsilon} \, \gamma^\beta \, - \, \gamma^\alpha \, \frac{-i\gamma q + m}{q^2 + m^2 - i\epsilon} \, \gamma^\beta \, .$$

We have attempted to show this graphically.

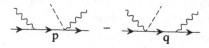

No factor is any longer associated with the dashed line, but we must nevertheless show it as there is still momentum k flowing through that line into the diagram. To show that the momentum k is no more there as a factor we removed the short double piece.

At this point it is reasonably clear how things work. Take together all diagrams where the dashed line is connected to an electron line in all possible ways and apply the above identity. Diagrams of succesive terms will cancel. Of course, if the dashed line has an external line either left or right there is directly zero there. Let us show this explicitly on a case with three photon lines. Thus consider, for a given configuration of all other lines, the diagrams with the dashed line in all possible positions along an electron line.

There are three more diagrams contributing in this case (exchange of the two other photon lines), but they form a group by themselves. Applying Ward's identity to the above set we obtain four non-zero diagrams.

We used that we get directly zero if an inverse propagator appears at an external line, and indicated that by showing a zero. This actually tells us that an inverse propagator is like the Dirac equation, which is of course no accident. Anyway, in the above graphical equation the first term cancels against the second and the third against the fourth, and the total result is zero.

If the electron line is a closed loop the same mechanism works, as shown in the figures. There are two ways to attach the line with the replaced polarization vector.

Applying Ward's identity gives four terms.

The first term cancels against the last, the second against the third.

Incidentally, there is another cancellation, the diagrams with the electron line with arrows in the opposite direction cancel against the diagrams shown. This is an application of Furry's theorem: diagrams with an odd number of external photon lines add up to zero. It is a consequence of behaviour under charge conjugation; the contribution of electrons cancels against the positron contribution if the number of vertices is odd, which implies a factor $(\text{charge})^n$ with odd n.

Exercise 9.1 Demonstrate Furry's theorem for a one loop electron graph with an odd number of photon lines attached. Thus show that reversal of the arrows leads to an overall minus sign. Use that the trace of an odd number of γ's is zero.

To recapitulate: the Ward identities for q.e.d. are in graphical form as shown in the figure. The dashed line with a short double piece is a photon line with the polarization vector replaced by its associated four-momentum. The blob stands

for all diagrams of a given order contributing to the process with the given external lines. There may be any number of external lines, in- or outgoing.

The external electron lines must be on mass shell and have the appropriate spinors $\bar{u}$ or u. The external photon lines need not to have a polarization vector, nor do they need to be on mass shell. We indicated that by terminating them with a cross. They may in fact contain the associated momentum k, or the space reflected $\tilde{k}$ as factor. That is important, because the identity is needed also for such cases. Conventionally the cross notation implies that there is also a propagator as factor, which is fine too.

It is actually no big deal to allow for electron lines off mass shell without spinors. Replace the spinor by a propagator; after applying Ward's identity there will be a left-over graph at the end (beginning) of the line. Now one must be careful about factors e etc. There is a factor $\pm e$ depending on the direction of the arrow. Again we use a cross to denote the absence of any special terminating conditions, and the factor just mentioned is part of the definition of the connection of the dashed line to a cross: $+e$ if the arrow of the electron line points towards the cross, else $-e$. Here it is important: to avoid inverse propagator factors on the right hand side, propagators must be associated with the electron lines connecting to a cross on the left hand side. If you want to be clear about all factors check the identity for the case of just one vertex in the blob on the left.

In this form the Ward identity is already quite close to the form of the Ward identities for non-abelian gauge theories. We repeat that the blobs must contain all diagrams of a given order in perturbation theory with a given set of external lines. There may be subclasses of diagrams for which the identity holds, but one must be very careful about that.

Appendix A
Complex Spaces, Matrices, CBH Equation

A.1 Basics

In this appendix some basic properties are reviewed. We will truly start from the bottom up, some might think even too far down.

Let there be given an n-dimensional space with a set of vectors $e_1, \ldots, e_n$, orthonormal with respect to some dot-product. Many equations in this appendix remain valid also for infinite n, sometimes subject to the convergence of a summation. Every vector in this space can be written as a linear combination of these basis vectors, with possibly complex coefficients. Consider two vectors a and b:

$$a = \alpha_1 e_1 + \alpha_2 e_2 + \cdots + \alpha_n e_n \qquad b = \beta_1 e_1 + \beta_2 e_2 + \cdots + \beta_n e_n.$$

Thus the α_i and β_i are (possibly complex) numbers. If λ is some (possibly complex) number then the vector λa is defined as a vector with components equal to those of a multiplied with λ. The components of the vector $a + b$ are the sum of those of a and b. The dot-product between a and b is:

$$(a, b) = \alpha_1^* \beta_1 + \alpha_2^* \beta_2 + \cdots + \alpha_n^* \beta_n.$$

Note that $(a, b) = (b, a)^*$, where * as usual implies complex conjugation. The following properties hold (λ some number):

$$(a, a) \geq 0, \quad (a, b + c) = (a, b) + (a, c),$$

$$(a, \lambda b) = \lambda(a, b), \quad (\lambda a, b) = \lambda^*(a, b)$$

Let there be given an $n \times n$ matrix A with matrix elements a_{ij}:

$$A = \begin{pmatrix} a_{11} & a_{12} & \cdots & a_{1n} \\ a_{21} & a_{22} & \cdots & a_{2n} \\ \cdots & & & \\ \cdots & & & \\ a_{n1} & a_{n2} & \cdots & a_{nn} \end{pmatrix}$$

The matrix-elements a_{ij} are (possibly complex) numbers. Let there furthermore be given an $n \times n$ matrix B with elements b_{ij}.

• Definition. The product of the matrices A and B is a matrix (called C here) with matrix elements given by:

$$c_{ij} = \sum_{k=1}^{n} a_{ik}b_{kj}. \tag{A.1}$$

As a matter of notation one writes

$$C = AB. \tag{A.2}$$

(A.1) defines the matrix-elements c given the a and b. Since the a and b are commuting numbers, the order in which they are written is irrelevant; for example we could as well have written

$$c_{ij} = \sum_{k=1}^{n} b_{kj}a_{ik}.$$

Equation (A.2) is a symbolic notation for (A.1). In general $C = AB \neq BA$, the matrix $C' = BA$ is a matrix with matrix-elements c'_{ij} given by

$$c'_{ij} = \sum_{k=1}^{n} b_{ik}a_{kj}.$$

In other words, matrix-elements commute, but not matrices.

The unit matrix I is a matrix with all elements except the diagonal elements equal to zero, while all diagonal elements are 1. Clearly $AI = IA = A$ for any matrix A.

Any matrix A defines a mapping of the n-dimensional space in itself, any vector a can be mapped into another vector a' with components α'_i:

$$\alpha'_i = \sum_{k} a_{ik}\alpha_k, \qquad \text{notation} \qquad a' = Aa.$$

The following important observation can be made.

• If $(a, Aa) = (a, a)$ for all complex vectors a then $A = I$.

The assertion can be demonstrated simply in steps. First take for a all possible basis vectors. Then it follows that all diagonal elements of A must be one. Next take the combinations $e_j + ie_k$ and $e_j + e_k$, with $j \neq k$. It follows then that $a_{jk} = a_{kj} = 0$.

Interestingly, the statement is not true in real spaces, and the condition $(a, Aa) = (a, a)$ for all real a implies only that $A = I + B$ where B is some anti-symmetrical matrix (i.e., $b_{ij} = -b_{ji}$).

Let $A^\dagger$ be the matrix obtained from A by reflection and complex conjugation. Thus $a^\dagger_{ij} = a^*_{ji}$. The equation $(Aa, b) = (a, A^\dagger b)$ may be verified:

$$(Aa, b) = \sum_{i,k}(a_{ik}\alpha_k)^*\beta_i = \sum_{i,k}(\alpha_k)^*a^\dagger_{ki}\beta_i = (a, A^\dagger b).$$

Let now U be a matrix such that every dot-product is invariant, i.e., $(a, b) = (Ua, Ub)$ for all vectors a and b. Such matrices are called unitary. The condition that U must satisfy in order to be unitary follows: $(a, b) = (Ua, Ub) = (a, U^\dagger U b)$ for all a and b. The conclusion is that $U^\dagger U = I$. Note, incidentally, that the invariance $(Ua, Ua) = (a, a)$ for all complex a suffices to deduce unitarity.

The foregoing equations are used to prove unitarity of the S-matrix as a consequence of conservation of probability.

A matrix A is called hermitian if $A = A^\dagger$. Then all eigenvalues of A are real and vice versa, which can be deduced trivially by considering $(a, Aa) = (a, A^\dagger a)$ with a any eigenvector of A.

• Definition. The commutator of two matrices A and B is given by

$$[A, B] = AB - BA. \tag{A.3}$$

If the matrices A and B commute the commutator is zero.

• The sum and difference respectively of two matrices A and B is a matrix (called C here) with matrix-elements given by:

$$c_{ij} = a_{ij} + b_{ij} \quad \text{and} \quad c_{ij} = a_{ij} - b_{ij}$$

respectively. Notation:

$$C = A + B \quad \text{and} \quad C = A - B$$

respectively. Obviously $A + B = B + A$.

• The product of a (possibly complex) number x and a matrix A is a matrix B with matrix-elements:

$$b_{ij} = x \cdot a_{ij}.$$

Notation: $B = xA$. It follows for example that $B^2 = x^2 A^2$ since $\sum_k b_{ik}b_{kj} = x^2 \sum_k a_{ik}a_{kj}$.

• Definition. The determinant of a matrix A (notation $Det(A)$) is a (possibly complex) number given by:

$$Det(A) = \sum_{k=1}^{n} a_{ik}(-1)^{k+i}Det(A(i,k)),$$

for any i. One speaks of developing with respect to row i. The $(n-1) \times (n-1)$ matrix $A(i,k)$ is obtained from the matrix A by omitting row i and column k. The determinant of a 1×1 matrix with element a is defined to be a.

An alternative definition is as follows. Let $\epsilon(i_1, i_2, \ldots, i_n)$ be the completely anti-symmetric tensor, anti-symmetric under exchange of any two indices, and with further $\epsilon(1, 2, \ldots, n) = 1$. Then:

$$a_{j_1 i_1} a_{j_2 i_2} \cdots a_{j_n i_n} \epsilon(i_1, i_2, \ldots, i_n) = Det(A)\epsilon(j_1, j_2, \ldots, j_n).$$

There is a summation over i_1, i_2, etc., which we did not show explicitly. In fact we follow the Einstein convention that every repeated index in a term implies a summation.

The equivalence of the two definitions is easy to show, by induction. Start by observing that the left hand side is obviously anti-symmetric for exchange of any two j, which tells you that the expression must be proportional to $\epsilon(j_1, j_2, \ldots, j_n)$, as shown. So we need to compute it only for $j_1 = 1$, $j_2 = 2$, etc. Write the summation over i_1 (or i_2 or whatever) explicitly. That gives then precisely the first definition of the determinant, development with respect to row 1 (or row 2 or whatever), assuming the validity of the equation for a matrix with one less row and column. This uses that the ϵ tensor with one index kept fixed is (apart from possibly a sign) the completely anti-symmmetric tensor in a space of one less dimension.

Multiplying the above equation with $\epsilon(j_1, j_2, \ldots, j_n)$ and summing over $j_1, j_2, \ldots j_n$ gives:

$$\epsilon(i_1, i_2, \ldots, i_n)\epsilon(j_1, j_2, \ldots, j_n)a_{j_1 i_1}a_{j_2 i_2}\cdots a_{j_n i_n} = n!Det(A)$$

This shows clearly that the determinant changes sign when exchanging two rows (exchanging two j) or two columns (exchanging two i), and that the determinant of the reflected of a matrix equals the determinant of the matrix (exchange of all the i and j). Several other properties of determinants are simple to deduce. For example, if $C = xA$ with x a number and A an $n \times n$ matrix then obviously $Det(C) = Det(xA) = x^n Det(A)$. Also the rule $Det(AB) = Det(A) \cdot Det(B)$ is not difficult to derive.

The inverse of a matrix A is a matrix B such that $BA = AB = I$. Notation: $B = A^{-1}$. The matrix-elements are:

$$b_{ij} = \frac{(-1)^{i+j} Det(A(j,i))}{Det(A)}.$$

Exercise A.1 Verify this for the special case of a 2×2 matrix A.

Note $A(j,i)$, not $A(i,j)$. It follows that a matrix A has an inverse if $Det(A) \neq 0$. The proof that $BA = I$ is simple if one realizes that $\sum_j b_{ij} a_{jk}$ is precisely computing a determinant, and one must only figure out what determinant. For $i = k$ that is then the determinant of A itself, for $i \neq k$ it is the determinant of a matrix with two equal rows, which is zero.

We close this section with some remarks concerning traces. The trace of a matrix A is defined as the sum of the diagonal elements:

$$\mathrm{Tr}(A) = \sum_i a_{ii}.$$

The trace of a product of matrices is invariant under cyclic rotation. For example:

$$\mathrm{Tr}(ABC) = \sum_{i,j,k} a_{ij} b_{jk} c_{ki} = \sum_{i,j,k} c_{ki} a_{ij} b_{jk} = \mathrm{Tr}(CAB).$$

A.2 Differentiation of Matrices

Matrix-elements may be functions of one or more variables. These functions may be differentiable, in which case the derivative of the matrix may be defined.

• Definition. The derivative of a matrix A is a matrix whose elements are the derivative of the matrix-elements of A.

$$A' = \frac{dA(x)}{dx}, \quad \text{matrix-elements} \quad a'_{ij} = \frac{da_{ij}(x)}{dx}.$$

The derivative of a matrix will in general not commute with the matrix itself. Apart from this the usual equations from differential calculus are valid here too, as for example $d(A+B)/dx = dA/dx + dB/dx$. However,

$$\frac{d(A^2)}{dx} = \frac{dA}{dx} \cdot A + A \cdot \frac{dA}{dx}.$$

This is a consequence of the chain rule:

$$\frac{d}{dx} \sum_k a_{ik} a_{kj} = \sum_k a'_{ik} a_{kj} + \sum_k a_{ik} a'_{kj}.$$

In general:

$$\frac{dAB}{dx} = A'B + AB'.$$

The derivative of the inverse of a matrix A can be found as follows. One has $A^{-1}A = I$ and thus

$$0 = \frac{d}{dx}(A^{-1}A) = \frac{dA^{-1}}{dx}A + A^{-1}\frac{dA}{dx},$$

or

$$\frac{dA^{-1}}{dx} = -A^{-1}\frac{dA}{dx}A^{-1}.$$

A.3 Functions of Matrices

Functions based on addition and multiplication may be defined for matrices as well. Especially the function e^x is important. There are two equivalent definitions:

$$e^A = \sum_{k=0}^{\infty} \frac{1}{k!} A^k, \quad \text{with } A^0 = I \tag{A.4}$$

$$e^A = \lim_{m \to \infty} (I + \frac{1}{m}A)^m. \tag{A.5}$$

The demonstration that the second definition is the same as the first is as usual, namely by application of the binomial expansion to the second equation.

Equation (A.5) allows the derivation of an interesting equation. Consider

$$Det(e^A) = \lim_{m \to \infty} \left[Det(I + \frac{1}{m}A) \right]^m. \tag{A.6}$$

The determinant inside the square brackets may be computed,

ignoring terms of order $1/m^2$ or smaller. Consider for example a 2×2 matrix. Then:

$$I + \frac{1}{m}A = \begin{pmatrix} 1 + \frac{a_{11}}{m} & \frac{a_{12}}{m} \\ \frac{a_{21}}{m} & 1 + \frac{a_{22}}{m} \end{pmatrix}.$$

If terms $1/m^2$ and higher can be ignored then in the calculation of the determinant the non-diagonal terms can be ignored, and the determinant is simply the product of the diagonal elements:

$$Det(I + \frac{1}{m}A) = (1 + \frac{a_{11}}{m})(1 + \frac{a_{22}}{m}) + \mathcal{O}(\frac{1}{m^2})$$

$$= 1 + \frac{1}{m}(a_{11} + a_{22}) + \mathcal{O}(\frac{1}{m^2}).$$

In general one finds:

$$Det(I + \frac{1}{m}A) = 1 + \frac{1}{m}\text{Tr}(A) + \mathcal{O}(\frac{1}{m^2}), \qquad \text{Tr}(A) = \sum_k a_{kk}.$$

$\text{Tr}(A)$ is called the trace of A, and is the sum of all elements along the diagonal. Inserting this in (A.6):

$$Det(e^A) = \lim_{m \to \infty} \left[1 + \frac{1}{m}\text{Tr}(A) + \mathcal{O}(\frac{1}{m^2}) \right]^m = e^{\text{Tr}(A)}. \quad (A.7)$$

That terms of order $1/m^2$ can be ignored may be established easily.

Exercise A.2 Show that

$$\lim_{m \to \infty} (1 + \frac{y}{m^2})^m = \text{Constant}$$

by differentiation with respect to y. The constant is obviously one.

As a preliminary to the work needed for the CBH equation we will derive a simple but useful equation. Consider the matrix $H(y)$ defined by

$$H(y) = e^{-yF} G e^{yF},$$

where F is not dependent on y. The quantity G is also independent of y, but apart from that may be anything, a matrix, or a constant, or as we will put later, an operator such as d/dx. Now compute the derivative of H with respect to y:

$$\frac{dH(y)}{dy} = \frac{de^{-yF}}{dy} G e^{yF} + e^{-yF} G \frac{de^{yF}}{dy}.$$

Using

$$\frac{d}{dy}e^{yF} = F \cdot e^{yF} = e^{yF} \cdot F,$$

which is rather obvious from (A.4) as F is no function of y, one finds:

$$\frac{dH(y)}{dy} = -e^{-yF}FGe^{yF} + e^{-yF}GFe^{yF} = e^{-yF}[G,F]e^{yF}.$$

Integrating this equation from 0 to 1:

$$H(1) - H(0) = e^{-F}Ge^F - G = \int_0^1 e^{-yF}[G,F]e^{yF}dy,$$

or, after multiplication with e^F

$$Ge^F - e^FG = \left[G, e^F\right] = \int_0^1 e^{(1-y)F}[G,F]e^{yF}dy.$$

Let now F be a function of some variable x, and take $G = d/dx$. If f and g are functions of x then clearly

$$(\frac{df}{dx})g = (\frac{d}{dx}f - f\frac{d}{dx})g = \left[\frac{d}{dx}, f\right]g.$$

In other words, $[d/dx, f] = df/dx$. Thus:

$$\frac{de^F}{dx} = \int_0^1 e^{(1-y)F}\frac{dF}{dx}e^{yF}dy.$$

A.4 The CBH Equation

The Campbell–Baker–Hausdorff equation is of the utmost importance in relation to the theory of Lie-groups. Examples of such groups relevant to physics are the Lorentz group and the various unitary groups governing the standard model and low energy hadron physics (isospin), namely SU(2) and SU(3). The CBH equation explains the central role of structure constants in the theory.

Consider the product of exponentials of two matrices A and B. The result can be written as the exponential of another matrix, C, and the CBH equation gives C in terms of A and B. Thus, let

$$e^C = e^A \cdot e^B$$

The CBH equation is:

$$C = A + B + \frac{1}{2}[A,B] + \text{ multiple commutators of } A \text{ and } B. \quad \text{(A.8)}$$

Multiple commutators are expressions of the type

$$[[A, B], B] = [A, B] \cdot B - B \cdot [A, B] = ABB - 2BAB + BBA$$
$$[[B, [A, B]], B]; \quad [[[A, B], B], B]; \quad \text{etc.}$$

The numerical coefficients of the multiple commutators are unfortunately difficult to establish in general, but for our purposes it is sufficient to understand that C is of the form (A.8), and this is what we will show here. Also the first few coefficients will be calculated.

Introduce a variable x, and study

$$e^{xA} e^{xB} = e^{C(x)}. \tag{A.9}$$

While the x-dependence of the left hand side is explicit in this equation, the x-dependence on the right hand side is complicated. The matrix C will be a complicated function of x. A series expansion of $C(x)$ with respect to x will be an expansion with respect to the number of factors A and B. Thus we must try to establish the functional dependence of C on x. Before doing that a few other equations must be derived. Consider the matrix H already introduced in the previous section:

$$H(y) = e^{-yF} G e^{yF}.$$

The matrix F is independent of the variable y, as is G. The matrix H can be written as a series in y:

$$H = H_0 + yH_1 + \frac{y^2}{2!} H_2 + \frac{y^3}{3!} H_3 + \cdots,$$

with

$$H_n = \frac{d^n H(y)}{dy^n} \Big|_{y=0}$$

The H_n must be computed. Differentiation of $H(y)$ with respect to y gives:

$$\frac{dH(y)}{dy} = \frac{de^{-yF}}{dy} G e^{-yF} + e^{-yF} G \frac{de^{yF}}{dy}.$$

Using $d\,exp(yF)/dy = F \cdot exp(yF) = exp(yF) \cdot F$ the result is:

$$\frac{dH(y)}{dy} = e^{-yF} [G, F] e^{yF}.$$

Similarly:

$$\frac{d^2 H(y)}{dy^2} = e^{-yF}[[G,F],F]e^{yF}$$

$$\frac{d^n H(y)}{dy^n} = e^{-yF}[[[\cdots[G,F],F],F]\cdots F]e^{yF}.$$

For $y = 0$:

$$H_n = [[[\cdots[G,F],F],F],\cdots F].$$

Inserting this into the equation for $H(y)$ we obtain:

$$e^{-yF}Ge^{yF} = G + y[G,F] + \frac{y^2}{2!}[[G,F],F] + \frac{y^3}{3!}[[[G,F],F],F] + \cdots$$

$$(A.10)$$

Now take F to be a function of the variable x and take for G the operator d/dx. Specializing furthermore to the case $y = 1$:

$$e^{-F(x)}\frac{d}{dx}e^{F(x)} = F' + \frac{1}{2!}[F',F] + \frac{1}{3!}[[F',F],F] + \cdots$$

$$\equiv <F',F>. \qquad (A.11)$$

The second line defines the notation $<F',F>$, with F' the derivative of the x-dependent matrix F.

Armed with equations (A.10) and (A.11) we go back to the beginning, eq (A.9), calculating the matrix $C(x)$. Differentiation of equation (A.9) gives:

$$\frac{d}{dx}e^{xA}e^{xB} = e^{xA}Ae^{xB} + e^{xA}Be^{xB}$$

$$= e^{xA}e^{xB}\left(e^{-xB}Ae^{xB} + e^{-xB}Be^{xB}\right)$$

$$= e^{C(x)} <C',C>.$$

Consider now the last two lines. Using equation (A.9) itself the outside factors cancel. Since B commutes with $exp(xB)$ the second term in between parentheses simplifies to B. For the first term in parentheses we may use equation (A.10), substituting x for y. This produces the principal result:

$$C' + \frac{1}{2!}[C',C] + \frac{1}{3!}[[C',C],C] + \cdots$$

$$= A + B + x[A,B] + \frac{x^2}{2!}[[A,B],B] + \frac{x^3}{3!}[[[A,B],B],B] + \cdots \qquad (A.12)$$

Writing $C(x)$ as a power series in x this equation may be used to solve for the coefficients of that expansion. Thus write:

$$C = xC_1 + x^2C_2 + x^3C_3 + \cdots$$

and by differentiation

$$C' = C_1 + 2xC_2 + 3x^2C_3 + \cdots$$

We used the rather ovious fact (from the starting equation (A.9)) that C is zero if x is zero. Inserting this into equation (A.12) and comparing powers of x one may find the C_i. From the form of the right hand side of equation (A.12) it follows that all these coefficients will depend only on multiple commutators. To make this explicit we will compute C_1, C_2 and C_3.

Keeping terms up to order x^2 the left hand side of equation (A.12) is (note that C contains at least one x so that terms containing three or more C can be neglected):

$$C' + \tfrac{1}{2}[C', C] + \tfrac{1}{6}[[C', C]C]$$

$$= C_1 + 2xC_2 + 3x^2C_3 + \frac{x^2}{2}[C_1, C_2] + x^2[C_2, C_1].$$

The triple commutator gives no x^2 terms. Comparing this with the right hand side of equation (A.12) gives immediately

$$C_1 = A + B, \qquad C_2 = \tfrac{1}{2}[A, B].$$

Inserting this the coefficient C_3 may be found:

$$C_3 = \tfrac{1}{12}[A, [A, B]] + \tfrac{1}{12}[[A, B], B].$$

In summary, for x equals one:

$$e^A e^B = e^C, \quad C = A + B + \tfrac{1}{2}[A, B] + \tfrac{1}{12}[A, [A, B]] + \tfrac{1}{12}[[A, B], B] + \cdots$$

where the dots stand for terms containing multiple commutators involving at least four A and B. Here the fourth and fifth order terms:

$$\tfrac{1}{48}[A, [[A, B], B]] + \tfrac{1}{48}[[A, [A, B]], B] + \tfrac{1}{120}[[A, [[A, B], B]], B]$$

$$+ \tfrac{1}{120}[A, [[A, [A, B]], B]]$$

$$- \tfrac{1}{360}[A, [[[A, B], B], B]] - \tfrac{1}{360}[[A, [A, [A, B]]], B]$$

$$- \tfrac{1}{720}[A, [A, [A, [A, B]]]] - \tfrac{1}{720}[[[[A, B], B], B], B].$$

Exercise A.3 Verify the fourth order terms.

Appendix B
Traces

B.1 General

The fundamental property of the γ-matrices, on which the calculation of the traces is based, is the anticommutation rule:

$$\{\gamma^\mu, \gamma^\nu\} \equiv \gamma^\mu\gamma^\nu + \gamma^\nu\gamma^\mu = 2\delta_{\mu\nu}\, I, \qquad (B.1)$$

where I is the unit matrix. Furthermore we will use:

$$\text{Trace}\,(I) \equiv [I] = 4. \qquad (B.2)$$

Here, and in the following, square brackets around the γ-matrices denote a trace. Consider now the trace of a product of an even number of γ-matrices:

$$[\gamma^{\mu_1}\gamma^{\mu_2}\ldots\gamma^{\mu_{2m}}].$$

Using the anticommutation rule (B.1) we have:

$$[\gamma^{\mu_1}\gamma^{\mu_2}\ldots\gamma^{\mu_{2m}}] = -\,[\gamma^{\mu_2}\gamma^{\mu_1}\ldots\gamma^{\mu_{2m}}]$$
$$+\, 2\delta_{\mu_1\mu_2}\,[\gamma^{\mu_3}\gamma^{\mu_4}\ldots\gamma^{\mu_{2m}}].$$

Going on in this way we may move γ^{μ_1} to the extreme right. The result is:

$$[\gamma^{\mu_1}\ldots\gamma^{\mu_{2m}}] = -\,[\gamma^{\mu_2}\ldots\gamma^{\mu_{2m}}\gamma^{\mu_1}]$$
$$+\, 2\sum_{i=2}^{2m}(-1)^i\delta_{\mu_1\mu_i}\,[\gamma^{\mu_2}\ldots\gamma^{\mu_{i-1}}\gamma^{\mu_{i+1}}\ldots\gamma^{\mu_{2m}}].$$

The trace of a product of matrices is unchanged under a cyclic rotation of those matrices. Thus the first term on the right hand side is equal to the initial term, on the left side. We get:

$$[\gamma^{\mu_1}\ldots\gamma^{\mu_{2m}}] = \sum_{i=2}^{2m}(-1)^i\delta_{\mu_1\mu_i}\,[\gamma^{\mu_2}\ldots\gamma^{\mu_{i-1}}\gamma^{\mu_{i+1}}\ldots\gamma^{\mu_{2m}}].$$

$$(B.3)$$

In this way the trace of a product of $2m$ γ-matrices can be expressed in traces of products of $2m - 2$ matrices. With (B.2) we have then determined the trace of any even product of γ-matrices.

Consider now the trace of a product of three γ's. With $\mu \neq \nu$:

$$[\gamma^\nu \gamma^\mu \gamma^\nu] = -[\gamma^\mu \gamma^\nu \gamma^\nu] = -[\gamma^\nu \gamma^\mu \gamma^\nu]. \qquad \text{(B.4)}$$

No summation over ν is intended. The first step is made by using the anticommutation rules, the second by rotation. We conclude from (B.4):

$$[\gamma^\nu \gamma^\mu \gamma^\nu] = 0 \qquad \text{(no summation over } \nu, \ \mu \neq \nu\text{)}.$$

Rotating the γ-matrices and using $\gamma^\nu \gamma^\nu = 1$ (this follows from the anticommutation rule (B.1)) we find:

$$[\gamma^\mu] = 0.$$

Up to now we have not used all the γ-matrices are involved, i.e., the range of the indices μ and ν in (B.1). Thus the equations derived up to now hold for an arbitrary number of γ's as long as these γ's satisfy the anticommutation rules (B.1).

We will now assume that we are dealing with a set of an even number of γ-matrices. This includes the case of interest, namely four γ-matrices. Then, as we will show now, it follows from (B.1) that the trace of a product of an odd number of γ-matrices is zero.

Consider the trace of $2m + 1$ γ-matrices. First we remove all γ's that occur twice, simply by anticommutation. For example:

$$\left[\gamma^3 \gamma^1 \gamma^2 \gamma^1 \gamma^4 \ldots\right] = -\left[\gamma^3 \gamma^2 \gamma^4 \ldots\right].$$

After this operation we are left with a trace of an odd number of γ's, with no two equal γ's. Since there is all together an even number of γ's there must be one γ-matrix that does not occur in that trace. Suppose this is γ^1, and let us denote the product of γ's in the trace by S. Since S contains an odd number of γ's of which none equal to γ^1 we have:

$$\gamma^1 S = -S\gamma^1.$$

Thus:

$$[S] = \left[S\gamma^1 \gamma^1\right] = \text{(cyclic rotation)} = \left[\gamma^1 S\gamma^1\right]$$

$$= \text{(anticommutation)} = -\left[S\gamma^1 \gamma^1\right] = -[S].$$

It follows that $[S] = 0$.

If the number of different γ-matrices is odd then the above proof fails. There remains one trace that can be non-zero, namely the trace of the product of all the γ-matrices. The commutation rules (B.1) are not sufficient to conclude anything for that trace. This may be illustrated for the case as encountered, namely the usual four γ-matrices as given explicitly before. Define further:

$$\gamma^5 = \gamma^1\gamma^2\gamma^3\gamma^4. \tag{B.5}$$

As also shown before one finds:

$$\gamma^5 = \begin{pmatrix} 0 & 0 & -1 & 0 \\ 0 & 0 & 0 & -1 \\ -1 & 0 & 0 & 0 \\ 0 & -1 & 0 & 0 \end{pmatrix}. \tag{B.6}$$

From (B.5) we find, using the anticommutation rules:

$$\gamma^\mu\gamma^5 + \gamma^5\gamma^\mu = 0 \qquad \mu = 1,2,3,4$$
$$\gamma^5\gamma^5 = 1.$$

Thus we now have a fifth γ-matrix, and the anticommutation rules (B.1) hold for $\mu,\nu = 1,\ldots 5$. All equations derived from (B.1) hold, for instance:

$$\left[\gamma^5\right] = 0$$
$$\left[\gamma^\mu\gamma^\nu\gamma^5\right] = 0 \qquad (\mu,\nu = 1,\ldots 4).$$

As long as not <u>all</u> γ's occur in the trace one gets zero for odd numbered traces. However,

$$\left[\gamma^1\gamma^2\gamma^3\gamma^4\gamma^5\right] = 4$$

because this is simply the trace of γ^5 squared, which is the unit matrix. It is easy to see that

$$\left[\gamma^\mu\gamma^\nu\gamma^\alpha\gamma^\beta\gamma^5\right] = 4\epsilon_{\mu\nu\alpha\beta} \qquad (\mu,\ldots\beta = 1,\ldots 4)$$

where $\epsilon_{\mu\nu\alpha\beta}$ is a tensor that is zero if two indices are equal, and moreover it changes sign if two indices are interchanged.

Finally we will derive a number of equations that hold only for the specific four-dimensional case as encountered in quantum electrodynamics. Thus we have the four γ-matrices as defined earlier, and γ^5 as given by (B.5), (B.6). In the following we will not really need the explicit form of the γ-matrices, but we use that they are 4×4 matrices.

From the four γ's and γ^5 we may form the following 16 independent 4×4 matrices:

$$
\begin{aligned}
&I \\
&\gamma^5 \\
&\gamma^\mu && \mu = 1, \ldots 4 \\
&\gamma^\mu \gamma^5 && \mu = 1, \ldots 4 \\
&\sigma^{\mu\nu} && \mu, \nu = 1, \ldots 4, \quad \mu < \nu
\end{aligned}
\tag{B.7}
$$

σ is defined as before:

$$
\sigma^{\mu\nu} = -\sigma^{\nu\mu} = \tfrac{1}{4}\left(\gamma^\mu\gamma^\nu - \gamma^\nu\gamma^\mu\right).
$$

To show that they are independent rests on the observation that the trace of the product of two different members of the above set is always zero, but not the trace of any one squared. Thus for example:

$$
\left[\sigma^{\mu\nu}\gamma^\alpha\gamma^5\right] = 0 \; ; \quad [\gamma^\mu\gamma^\nu] = 4\delta_{\mu\nu}.
$$

Another example:

$$
\left[\sigma^{\mu\nu}\sigma^{\alpha\beta}\right] = -\delta_{\mu\alpha}\delta_{\nu\beta} + \delta_{\mu\beta}\delta_{\nu\alpha}.
$$

Since we have 16 matrices we have a complete set if we can show that none can be expressed as a linear combination of the others. Suppose for instance that γ^1 is a linear combination of the rest:

$$
\gamma^1 = a_1 I + a_2\gamma^2 + a_3\gamma^3 + \cdots a_{34}\sigma^{34}.
$$

The equation is clearly wrong, because if we multiply both sides with γ^1 and take the trace we find $4 = 0$.

Any 4×4 matrix can therefore be expressed as a linear combination of the set (B.7). Let S be a product of γ-matrices. Since S is also a 4×4 matrix we have:

$$
S = a_0 I + a_5\gamma^5 + a_\mu\gamma^\mu + a_\mu^5\gamma^\mu\gamma^5 + a_{\mu\nu}\sigma^{\mu\nu}
\tag{B.8}
$$

with 16 coefficients $a_0 \ldots a_{34}$. The summation over μ and ν in the last term goes over all indices μ and ν, but we take $a_{\mu\nu} = -a_{\nu\mu}$. Multiplying eq. (B.8) left and right with the various members of the set (B.7) and taking the trace gives:

$$
a_0 = \tfrac{1}{4}[S]
$$
$$
a_5 = \tfrac{1}{4}\left[\gamma^5 S\right]
$$
$$
a_\mu = \tfrac{1}{4}\left[\gamma^\mu S\right]
$$

$$a_\mu^5 = \tfrac{1}{4}\left[\gamma^5\gamma^\mu S\right]$$
$$a_{\mu\nu} = \tfrac{1}{2}\left[\sigma^{\nu\mu}S\right].$$

Suppose now S contains an odd number of γ-matrices, whereby γ^5 must be counted as even, as shown by eq. (B.5). We then have $a_0 = a_5 = a_{\mu\nu} = 0$, and thus:

$$S^{\mathrm{odd}} = a_\mu\gamma^\mu + a_\mu^5\gamma^\mu\gamma^5.$$

Consider next an odd string S in between two γ's with equal indices, and with as usual summation intended. We have

$$\gamma^\alpha S^{\mathrm{odd}}\gamma^\alpha = a_\mu\gamma^\alpha\gamma^\mu\gamma^\alpha + a_\mu^5\gamma^\alpha\gamma^\mu\gamma^5\gamma^\alpha.$$

Now:

$$\gamma^\alpha\gamma^\mu\gamma^\alpha = -\gamma^\mu\gamma^\alpha\gamma^\alpha + 2\delta_{\mu\alpha}\gamma^\alpha$$
$$= -4\gamma^\mu + 2\gamma^\mu = -2\gamma^\mu$$

where the summation $\alpha = 1,\dots 4$ has been performed. Similarly

$$\gamma^\alpha\gamma^\mu\gamma^5\gamma^\alpha = 2\gamma^\mu\gamma^5.$$

We find:

$$\gamma^\alpha S^{\mathrm{odd}}\gamma^\alpha = -2a_\mu\gamma^\mu + 2a_\mu^5\gamma^\mu\gamma^5. \tag{B.9}$$

Let us now define S^R as the string obtained from S by reversing the order of all γ's. From the general treatment (B.2)–(B.3) we see that the trace of any product of γ's is unchanged if we reverse the order of all these γ's. We have:

$$[\gamma^\mu S] = \left[S^R\gamma^\mu\right] = \left]\gamma^\mu S^R\right]$$
$$\left[\gamma^5\gamma^\mu S\right] = \left[S^R\gamma^\mu\gamma^5\right] = \left[\gamma^\mu\gamma^5 S^R\right] = -\left[\gamma^5\gamma^\mu S^R\right].$$

Thus

$$a_\mu = \tfrac{1}{4}[\gamma^\mu S] = \tfrac{1}{4}\left[\gamma^\mu S^R\right]$$
$$a_\mu^5 = \tfrac{1}{4}\left[\gamma^5\gamma^\mu S\right] = -\tfrac{1}{4}\left[\gamma^5\gamma^\mu S^R\right]$$

and

$$S^R = \tfrac{1}{4}\left[\gamma^\mu S^R\right]\gamma^\mu + \tfrac{1}{4}\left[\gamma^5\gamma^\mu S^R\right]\gamma^\mu\gamma^5$$
$$= a_\mu\gamma^\mu - a_\mu^5\gamma^\mu\gamma^5. \tag{B.10}$$

We conclude from (B.9) and (B.10):

$$\gamma^\alpha S\gamma^\alpha = -2S^R \qquad (S \text{ odd}). \tag{B.11}$$

A similar equation can be obtained for even S if we take off one γ. With $S = \gamma^\lambda S'$, and S' odd we have:

$$\gamma^\alpha S \gamma^\alpha = \gamma^\alpha \gamma^\lambda S' \gamma^\alpha = -\gamma^\lambda \gamma^\alpha S' \gamma^\alpha + 2\delta_{\alpha\lambda} S' \gamma^\alpha$$
$$= 2\gamma^\lambda S'^R + 2S' \gamma^\lambda \qquad (S \text{ even}). \qquad (B.12)$$

The equations (B.11) and (B.12) are called **Chisholm identities**. Instead of (B.12) for an even string a more useful equation can be derived as follows. For even S we have:

$$S = a_0 I + a_5 \gamma^5 + a_{\mu\nu} \sigma^{\mu\nu}.$$

Now, by explicit calculation:

$$\gamma^\alpha I \gamma^\alpha = 4 \cdot I$$
$$\gamma^\alpha \gamma^5 \gamma^\alpha = -4\gamma^5$$
$$\gamma^\alpha \sigma^{\mu\nu} \gamma^\alpha = 0$$

and

$$\gamma^\alpha S \gamma^\alpha = [S]I - \left[\gamma^5 S\right] \gamma^5 \quad (S \text{ even}). \qquad (B.13)$$

Thus for even S the quantity $\gamma^\alpha S \gamma^\alpha$ is a linear combination of I and γ^5.

Note that in the above γ^5 must always be counted separately. The equations (B.11) and (B.13) can be used with great advantage to eliminate any doubly occurring γ, with summation over the index of that γ. There are also equations for the case that no summation is involved. Let $\not{p}$ denote a linear combination of γ's:

$$\not{p} = p_1 \gamma^1 + p_2 \gamma^2 + p_3 \gamma^3 + p_4 \gamma^4.$$

One may derive, by methods similar to the above:

$$\not{p} S \not{p} = -p^2 S^R + \tfrac{1}{2} \left[\not{p} S^R\right] \not{p} + \tfrac{1}{2} \left[\gamma^5 \not{p} S^R\right] \not{p} \gamma^5 \quad (S \text{ odd})$$
$$\not{p} S \not{p} = -p^2 S^R + \tfrac{1}{2} \left[\not{p} \gamma^\alpha S^R\right] \gamma^\alpha \not{p} \quad (S \text{ even})$$

where

$$p^2 = p_1^2 + p_2^2 + p_3^2 + p_4^2.$$

Another identity, in fact also due to Chisholm, can be obtained as follows. Let S be an odd string. We have:

$$S = \tfrac{1}{4} [\gamma^\mu S] \gamma^\mu + \tfrac{1}{4} \left[\gamma^5 \gamma^\mu S\right] \gamma^\mu \gamma^5 \quad (S \text{ odd})$$
$$S^R = \tfrac{1}{4} [\gamma^\mu S] \gamma^\mu - \tfrac{1}{4} \left[\gamma^5 \gamma^\mu S\right] \gamma^\mu \gamma^5$$

Summing both equations:

$$S + S^R = \tfrac{1}{2} \left[\gamma^\mu S\right] \gamma^\mu. \tag{B.14}$$

This may be used to reduce the product of two traces to the sum of two traces:

$$[A\gamma^\mu B]\,[\gamma^\mu S] = 2[A\,S\,B] + 2[A\,S^R\,B] \quad (S \text{ odd}) \tag{B.15}$$

where A and B are arbitrary strings of γ-matrices.

The equations (B.11), (B.13) and (B.14) can be used to eliminate all γ-pairs from a trace. For the general case of several pairs of γ's one has the Kahane algorithm, which is easy to use, easy to prove (by induction) but somewhat difficult to describe. It is valid only in the case of four-dimensional space-time. In that case we have γ-matrices that obey the expansion rule (B.8). For a proof and consideration of other than four-dimensional space-time we refer to the literature (P. Van Nieuwenhuizen, *Supergravity* 1981, Cambridge Univ. Press 1982, page 151, in particular pages 160 and following; M. Veltman, *Nucl. Phys.* B319 (1989) 253).

B.2 Multi-Dimensional γ-Matrices

We conclude this appendix by showing how one can construct larger sets of γ-matrices that obey the anticommutation rules (B.1).

Consider the Pauli spin matrices τ^i, $i = 1, 2, 3$:

$$\tau^1 = \begin{pmatrix} 0 & 1 \\ 1 & 0 \end{pmatrix} \qquad \tau^2 = \begin{pmatrix} 0 & -i \\ i & 0 \end{pmatrix} \qquad \tau^3 = \begin{pmatrix} 1 & 0 \\ 0 & -1 \end{pmatrix}.$$

The τ^i, $i = 1, 2, 3$ satisfy anticommutation rules:

$$\tau^i \tau^j = -\tau^j \tau^i \quad \text{if } i \neq j$$
$$\tau^i \tau^i = 1.$$

The matrices $\gamma^1 ... \gamma^4$ can be written as:

$$\gamma^\mu = \begin{pmatrix} 0 & -i\tau^\mu \\ i\tau^\mu & 0 \end{pmatrix} \qquad \mu = 1, 2, 3$$

$$\gamma^4 = \begin{pmatrix} I & 0 \\ 0 & -I \end{pmatrix} \qquad \gamma^5 = \begin{pmatrix} 0 & -I \\ -I & 0 \end{pmatrix}$$

where I is the 2×2 unit matrix. This may be written in evident notation as:

$$\gamma^\mu = \underline{\tau}^2 \times \tau^\mu \qquad \gamma^4 = \underline{\tau}^3 \times I \qquad \gamma^5 = -\underline{\tau}^1 \times I.$$

The anticommutation rules follow from the anticommutation rules for the τ. Underlined objects commute with non-underlined objects. One has for $\mu \neq 4$:

$$\gamma^\mu \gamma^4 = \underline{\tau}^2 \times \tau^\mu \cdot \underline{\tau}^3 \times I = -\underline{\tau}^3 \times I \cdot \underline{\tau}^2 \times \tau^\mu = -\gamma^4 \gamma^\mu.$$

A set of seven 8×8 γ-matrices can be constructed along the same lines:

$$\Gamma^\mu = \begin{pmatrix} 0 & -i\gamma^\mu \\ i\gamma^\mu & 0 \end{pmatrix} \qquad \mu = 1, ...5$$

$$\Gamma^6 = \begin{pmatrix} I & 0 \\ 0 & -I \end{pmatrix} \qquad \Gamma^7 = \begin{pmatrix} 0 & I \\ I & 0 \end{pmatrix}$$

or

$$\Gamma^\mu = \tau^2 \times \gamma^\mu \qquad \Gamma^6 = \tau^3 \times I \qquad \Gamma^7 = \tau^1 \times I.$$

Note that Γ^7 is proportional to the product $\Gamma^1 \Gamma^2 \Gamma^3 \Gamma^4 \Gamma^5 \Gamma^6$. Whether these Γ-matrices have anything to do with some representation of a higher dimensional Lorentz group is another question that will not be considered here.

B.3 Frequently Used Equations

Finally we write down the expressions for the most frequently encountered traces. When doing a trace, first get the γ^5's out of the way, moving them all to the right so that in the end at most one γ^5 occurs, and that at the far right. Try to keep the combinations $1 \pm \gamma^5$ together, in order to use $(1+\gamma^5)(1-\gamma^5) = 0$ and $(1 \pm \gamma^5)(1 \pm \gamma^5) = 2(1 \pm \gamma^5)$.

$$[\gamma^\mu \gamma^\nu] = 4\,\delta_{\mu\nu}$$
$$[\gamma^\mu \gamma^\nu \gamma^\alpha \gamma^\beta] = 4\,\delta_{\mu\nu}\delta_{\alpha\beta} + 4\,\delta_{\mu\beta}\delta_{\nu\alpha} - 4\,\delta_{\mu\alpha}\delta_{\nu\beta}$$
$$[\gamma^\mu \gamma^\nu \gamma^\alpha \gamma^\beta \gamma^5] = 4\,\epsilon_{\mu\nu\alpha\beta}$$

This last equation is limited to four dimensions. In practice, the equations most used to simplify are (B.11) and (B.13), in particular:

$$\gamma^\mu \gamma^\alpha \gamma^\mu = -2\gamma^\alpha \quad \text{and} \quad \gamma^\alpha \gamma^\mu \gamma^\nu \gamma^\alpha = 4\delta_{\mu\nu} I.$$

If there are loop integrals it may be necessary to use such equations also for dimensions other than four:

$$\gamma^\mu \gamma^\alpha \gamma^\mu = (2 - n)\gamma^\alpha \quad \text{and} \quad \gamma^\alpha \gamma^\mu \gamma^\nu \gamma^\alpha = 4\delta_{\mu\nu} \, I + (n - 4)\gamma^\mu \gamma^\nu$$

where n is the number of dimensions.

Appendix C
Dimensional Regularization

In addition to the more or less standard equations in relation to dimensional regularization we will also give a number of standard equations that we have found useful in various circumstances.

Polar coordinates in n dimensions:

$$\int d_n x \, f(x) =$$

$$\int f(x) \, r^{n-1} dr \, \sin^{n-2} \theta_{n-1} d\theta_{n-1} \, \sin^{n-3} \theta_{n-2} d\theta_{n-2} \ldots d\theta_1 .$$

The integration limits are $(0, \infty)$ for r, and $0 \leq \theta_i \leq \pi$, except for $i = 1$: $0 \leq \theta_1 \leq 2\pi$. When $f(x)$ has no angular dependence the integrations over the θ may be done. The θ_1 integral is simply 2π, for the others one may use:

$$\int_0^\pi \sin^m \theta \, d\theta = \sqrt{\pi} \, \frac{\Gamma\left(\frac{m+1}{2}\right)}{\Gamma\left(\frac{m+2}{2}\right)} .$$

Then:

$$\int d_n x \, f(r) = \frac{2\pi^{n/2}}{\Gamma\left(\frac{n}{2}\right)} \int f(r) \, r^{n-1} dr.$$

The Γ function satisfies $z\Gamma(z) = \Gamma(z+1)$. If the argument is near zero one may use:

$$\Gamma(z) = \frac{1}{z} - \gamma + \left(\frac{\gamma^2}{2} + \frac{\pi^2}{12}\right) z + \mathcal{O}(z^2).$$

where $\gamma = 0.577215665$ is Euler's constant. Often $f(r)$ is of the form $1/(r^2 + A)^\alpha$, and in that case the equation

$$\int_0^\infty \frac{r^\beta dr}{(r^2 + A)^\alpha} = \frac{1}{2} \frac{\Gamma\left(\frac{\beta+1}{2}\right) \Gamma\left(\alpha - \frac{\beta+1}{2}\right)}{A^{\alpha - (\beta+1)/2} \Gamma(\alpha)}$$

may be useful.

In doing integrals Feynman's equation for combining denominators is very useful:

$$\frac{1}{a^j b^k} = \frac{(j+k-1)!}{(j-1)!(k-1)!} \int_0^1 \frac{x^{j-1}(1-x)^{k-1}}{(ax+b(1-x))^{j+k}}\, dx.$$

The simplest application, both j and k equal to 1:

$$\int d_n p \frac{1}{(p^2+m^2)(p^2+M^2)} =$$

$$\int_0^1 dx \int d_n p \frac{1}{(p^2 + xm^2 + (1-x)M^2)^2}.$$

For one loop integrals the basic equation is:

$$\mathcal{F}_\alpha(k) \equiv \int d_n p \frac{1}{(p^2 + 2kp + m^2 - i\epsilon)^\alpha}$$

$$= \frac{i\pi^{n/2}}{(m^2 - k^2)^{\alpha - n/2}} \frac{\Gamma(\alpha - n/2)}{\Gamma(\alpha)}.$$

To do this integral first shift $p = p' - k$ and then apply the equations given before.

This equation is the starting point for the derivation of many others. By differentiation with respect to k_μ one obtains equations for this type of integral with factors p_μ in the numerator. For example, differentiating twice with respect to k and then setting $k = 0$ one obtains:

$$\lim_{k \to 0} \frac{\partial}{\partial k_\mu} \frac{\partial}{\partial k_\nu} \mathcal{F}_\alpha(k) = \int d_n p \frac{4\alpha(\alpha+1)p_\mu p_\nu}{(p^2 + m^2 - i\epsilon)^{\alpha+2}}$$

$$= \frac{i\pi^{n/2}(\alpha - n/2)2\delta_{\mu\nu}\Gamma(\alpha - n/2)}{(m^2)^{\alpha - n/2 + 1}\Gamma(\alpha)}.$$

From this it readily follows:

$$\int d_n p \frac{p_\mu p_\nu}{(p^2 + m^2)^\alpha} = \frac{1}{2(\alpha-1)}\mathcal{F}_{\alpha-1}(0)\,\delta_{\mu\nu}.$$

This relation will be used below. Also, it may be seen that integrals over a polynomial of p are zero, by taking α zero or negative (from $\Gamma(\alpha) = \infty$ if α integer ≤ 0):

$$\int d_n p\, (p^2)^j = 0 \text{ if } j \geq 0, j \text{ integer}.$$

We may warn the reader of a difficult, ambiguous case (in the region around $n = 4$):

$$\int d_n p \, \frac{1}{p^4} = \lim_{m \to 0} \lim_{\alpha \to 2} \int d_n p \, \frac{1}{(p^2 + m^2)^\alpha} \, .$$

The result of this integral depends on how the limits are taken. Thus more precise specification coming from the problem considered must be used before this integral can be established.

Usually, at the end, one must take the limit of $n = 4$ (or $d = 0$), which is at times cumbersome. Here follows a collection of equations selected on the basis that they have been useful at some time, either for direct calculations of one loop graphs, or in other contexts such as estimating behaviour with respect to some variable, requiring series expansion. First a few equations that everyone knows but are never there when you need them.

$$\ln(1 + \epsilon) = \epsilon - \frac{\epsilon^2}{2} + \frac{\epsilon^3}{3} - \cdots$$

$$\sqrt{1 + \epsilon} = 1 + \frac{\epsilon}{2} - \sum_{j=2}^{\infty} \frac{(-1)^j \, 1 \cdot 3 \cdot 5 \cdots (2j - 3)}{2^j \, j!} \epsilon^j$$

$$= 1 + \frac{\epsilon}{2} - \frac{\epsilon^2}{8} + \frac{\epsilon^3}{16} - \frac{5\epsilon^4}{128} + \frac{7\epsilon^5}{256} \cdots$$

A few notations are used. The δ-function with more than two indices:

$$\delta_{\mu_1 \mu_2 \cdots \mu_\lambda} = \sum_{j=2}^{\lambda} \delta_{\mu_1 \mu_j} \delta_{\mu_2 \cdots \mu_{j-1} \mu_{j+1} \cdots \mu_\lambda}.$$

For example,

$$\delta_{\mu\nu\alpha\beta} = \delta_{\mu\nu}\delta_{\alpha\beta} + \delta_{\mu\alpha}\delta_{\nu\beta} + \delta_{\mu\beta}\delta_{\nu\alpha}.$$

The quantity Δ contains the infinite part if that occurs:

$$\Delta \equiv \frac{2\pi^2}{n - 4} + \pi^2(\gamma + \ln(\pi)) - \pi^2 c, \quad c_{\overline{MS}} = \frac{2}{n - 4} + \gamma + \ln(\pi).$$

The constant c depends on the subtraction scheme chosen, the value shown involving Euler's constant is for the $\overline{MS}$ scheme.

Now define:

$$\int d_n p \, \frac{1}{(p^2 + m^2)^j} = F_j$$

$$\int d_n p \, \frac{p_\mu p_\nu}{(p^2 + m^2)^j} = \delta_{\mu\nu} \, G_j$$

$$\int d_n p \, \frac{p_\mu p_\nu p_\alpha p_\beta}{(p^2 + m^2)^j} = \delta_{\mu\nu\alpha\beta} \, H_j$$

$$\int d_n p \, \frac{p_\mu p_\nu p_\alpha p_\beta p_\gamma p_\lambda}{(p^2 + m^2)^j} = \delta_{\mu\nu\alpha\beta\gamma\lambda} \, I_j \, .$$

Then (further down, the function F_1 will be called $A(m)$):

$$F_1 = im^2 \Delta + i\pi^2 m^2 \ln(m^2) - i\pi^2 m^2 \, ,$$

$$F_2 = -i\Delta - i\pi^2 \ln(m^2) \, ,$$

$$F_j = \frac{i\pi^2}{(j-1)(j-2)(m^2)^{j-2}} \, , \quad j \geq 3 \, .$$

$$G_1 = -\frac{im^4}{4} \Delta - \frac{i\pi^2}{4} m^4 \ln(m^2) + \frac{3i\pi^2}{8} m^4 \, ,$$

$$G_j = \frac{1}{2(j-1)} F_{j-1} \, .$$

$$H_1 = \frac{im^6}{24} \Delta + \frac{i\pi^2}{24} m^6 \ln(m^2) - \frac{11i\pi^2}{144} m^6 \, ,$$

$$H_j = \frac{1}{2(j-1)} G_{j-1} \, .$$

$$I_1 = -\frac{im^8}{192} \Delta - \frac{i\pi^2}{192} m^8 \ln(m^2) + \frac{25i\pi^2}{2304} m^8 \, ,$$

$$I_j = \frac{1}{2(j-1)} H_{j-1} \, .$$

The relation between G_j and F_{j-1} was derived before. The relations between H and G etc. can be derived similarly. There is also a relation between the X_1, consider for example G_1. The technique used here is very useful in general, and is used further down also to express functions B_1, B_{12} and B_{22} in terms of B_0. By multiplication with $\delta_{\mu\nu}$ one obtains (using that $\delta_{\mu\mu} = n$, the number of space-time dimensions):

$$nG_1 = \int d_n p \, \frac{p^2}{p^2 + m^2} = \int d_n p \, \frac{p^2 + m^2 - m^2}{p^2 + m^2}$$

$$= -m^2 F_1 \, .$$

One must be careful with n, it cannot straightaway be put equal to 4 because of the occurrence of terms $1/(n-4)$ in the quantity Δ. Thus use

$$\frac{1}{n} = \frac{1}{4 + n - 4} \approx \frac{1}{4} - \frac{n-4}{16} \, .$$

Now the result for G_1 follows easily. In general one may proceed as follows. Denote the expression corresponding to an integral with κ momenta by $f(\kappa)$. For example, $f(4) = H_1$ and $f(6) = I_1$. Write:

$$f(\kappa) = a_\kappa \Delta + b_\kappa \ln(m^2) + c_\kappa .$$

Then

$$a_\kappa = -\frac{m^2}{\kappa + 2} a_{\kappa-2}, \quad b_\kappa = -\frac{m^2}{\kappa + 2} b_{\kappa-2},$$

$$c_\kappa = -\frac{m^2}{\kappa + 2} c_{\kappa-2} + \frac{2\pi^2 m^2}{(\kappa + 2)^2} a_{\kappa-2}.$$

A few equations similar to the above, but now with two different factors in the denominator. Define:

$$f_{jk} \equiv \int \frac{1}{(p^2 + m^2)^j (p^2 + M^2)^k} .$$

Then:

$$f_{11} = -i\Delta + i\pi^2 \left[1 + \frac{M^2}{m^2 - M^2} \ln(M^2) - \frac{m^2}{m^2 - M^2} \ln(m^2) \right]$$

$$f_{12} = i\pi^2 \left[\frac{1}{M^2 - m^2} - \frac{m^2}{(M^2 - m^2)^2} \ln\left(\frac{M^2}{m^2}\right) \right]$$

$$f_{22} = i\pi^2 \left[-\frac{2}{(M^2 - m^2)^2} + \frac{M^2 + m^2}{(M^2 - m^2)^3} \ln\left(\frac{M^2}{m^2}\right) \right]$$

$$f_{13} = \frac{i\pi^2}{2} \left[\frac{M^2 + m^2}{M^2(M^2 - m^2)^2} - \frac{2m^2}{(M^2 - m^2)^3} \ln\left(\frac{M^2}{m^2}\right) \right]$$

$$f_{23} = \frac{i\pi^2}{2} \left[-\frac{5M^2 + m^2}{M^2(M^2 - m^2)^3} + \frac{2M^2 + 4m^2}{(M^2 - m^2)^4} \ln\left(\frac{M^2}{m^2}\right) \right]$$

$$f_{14} = \frac{i\pi^2}{6} \left[\frac{2M^4 + 5M^2 m^2 - m^4}{M^4(M^2 - m^2)^3} - \frac{6m^2}{(M^2 - m^2)^4} \ln\left(\frac{M^2}{m^2}\right) \right]$$

$$\int d_n p \, \frac{p_\mu p_\nu; \quad p_\mu p_\nu p_\alpha p_\beta; \quad p_\mu p_\nu p_\alpha p_\beta p_\gamma p_\lambda}{(p^2 + m^2)(p^2 + M^2)} =$$

$$\delta_{\mu\nu} X; \quad \delta_{\mu\nu\alpha\beta} Y; \quad \delta_{\mu\nu\alpha\beta\gamma\lambda} Z$$

$$X = \frac{i\Delta}{4}(M^2 + m^2) - \frac{3i\pi^2}{8}(M^2 + m^2)$$
$$+ \frac{i\pi^2}{4(M^2 - m^2)}\left(M^4 \ln(M^2) - m^4 \ln(m^2)\right)$$

$$Y = \frac{-i\Delta - i\pi^2\left(\ln(M^2) - \frac{11}{6}\right)}{24}\left(m^4 + m^2 M^2 + M^4\right)$$
$$- \frac{i\pi^2 m^6}{24(M^2 - m^2)}\ln\left(\frac{M^2}{m^2}\right)$$

$$Z = \frac{i\Delta + i\pi^2\left(\ln(M^2) - \frac{25}{12}\right)}{192}\left(m^6 + M^2 m^4 + M^4 m^2 + M^6\right)$$
$$+ \frac{i\pi^2 m^8}{192(M^2 - m^2)}\ln\left(\frac{M^2}{m^2}\right).$$

Equations for expressions where the denominator factors to some power may be obtained by differentiation. The limit $M = m$ can be tricky, and one should directly use the earlier given expressions in such cases.

One loop self-energy diagrams usually involve the function B_0 defined by:

$$B_0(k, m, M) = \int d_n p \frac{1}{(p^2 + m^2 - i\epsilon)((p + k)^2 + M^2 - i\epsilon)}$$
$$= -i\Delta - i\pi^2 \int_0^1 dx \, \ln\chi.$$

The function χ is quadratic in x, and in doing the x-integral one must be very careful with the roots and the imaginary part of this expression:

$$\chi = x(1 - x)k^2 + xM^2 + (1 - x)m^2 - i\epsilon$$
$$= sx^2 + x(-s + M^2 - m^2) + m^2 - i\epsilon, \text{ with } s = -k^2.$$

The thing to do first is to analyse the argument of the logarithm, in order to find out where this argument can be negative. The logarithm will develop an imaginary part if the argument is negative. Then this imaginary part is $-i\pi$, because the $-i\epsilon$ tells us that the argument of the logarithm must be taken slightly below the real axis, to which corresponds an angle of $-\pi$ if we think of this argument in the form $ce^{i\varphi}$. First note the obvious fact that the value of the argument is m^2 for $x = 0$, and M^2 for $x = 1$, both positive values. Let now $s < 0$. Then for x very large, either positive or negative, the function becomes negative. It therefore follows that the function crosses the real axis somewhere between $x = -\infty$ and $x = 0$, and also between $x = 1$ and $x = \infty$. Consequently the function is positive for $0 \le x \le 1$, i.e., in the whole integration region. Thus the argument of the logarithm is always positive in the integration region if $s < 0$.

If $s > 0$ then the function becomes positive for $x = \pm\infty$. Since the function is also positive for $x = 0$ and $x = 1$ it follows that the intersections with the real axis (the roots of the quadratic expression χ) must either be both in the region $0 \le x \le 1$, or both outside. We are interested in the case that they are both inside, because then the region between these roots is where the argument will be negative, and the imaginary part of the logarithm equal to $-i\pi$. The imaginary part of the integral over this logarithm is then the length of this interval times $-i\pi$. For $s > 0$ the minimum of the quadratic curve is reached for $x = 1/2 + (m^2 - M^2)/2s$, and this minimum is somewhere between the two roots. We are therefore interested in the case that this minimum is in the region $0 \le x \le 1$, or $s > |m^2 - M^2|$.

Consider now the two roots of χ:

$$x_{1,2} = \frac{1}{2s}\left[s - M^2 + m^2 \pm \sqrt{(s - M^2 - m^2)^2 - 4m^2M^2}\right].$$

These roots exist if the argument of the square root is positive, that is if $s < (M - m)^2$ or $s > (M + m)^2$. According to the discussion above, the first s region will give roots outside the integration region. Thus, finally, we conclude that the two roots will be in the integration region if $s > (M + m)^2$. The length of the region is the difference between the two roots. We so find the real part of the function B_0:

$$\text{Real}(B_0) = -\frac{\pi^3}{s}\sqrt{(s - M^2 - m^2)^2 - 4m^2M^2}\,\theta(s - (M + m)^2).$$

The real part of the integral is trivial. One can take the absolute value of the argument of the logarithm. With $\chi = s(x - x_1)(x - x_2)$, and using that $\int \ln(x + a) = (x + a)\ln(x + a) - x$ we find after some algebra:

$$B_0(k, m, M) = -\frac{\pi^3 \Lambda}{s} \theta(s - (M + m)^2)$$

$$- i\Delta - i\pi^2 \left[R + \ln(Mm) + \frac{M^2 - m^2}{2s} \ln\left(\frac{M^2}{m^2}\right) - 2 \right].$$

Remember, $s = -k^2$. The quantity R is:

$$R = -\frac{\Lambda}{s} \arctan\left(\frac{\Lambda}{s - M^2 - m^2}\right) \quad \text{if } (M - m)^2 < s < (M + m)^2;$$

$$R = \frac{\Lambda}{s} \ln\left|\frac{s - M^2 - m^2 + \Lambda}{2Mm}\right| \quad \text{otherwise.}$$

In here Λ is given by:

$$\Lambda = \sqrt{|(s - M^2 - m^2)^2 - 4m^2 M^2|}.$$

The arctan in the above equation is such that the resulting angle is negative, i.e., in the region $(-\pi, 0)$. Quite obviously, the special cases $M = 0$, $m = 0$ and/or $s = 0$ require special consideration, which means going back to beginning. Some of these cases are shown below, the case $s = 0$ was shown before (the function f_{11}).

We will next give this function B_0 for a few special values of the parameters.

$$B_0(k, 0, M) = -i\Delta + i\pi^2 \left[2 - \ln(M^2) \right.$$

$$\left. - \frac{k^2 + M^2}{k^2} \ln\left(\frac{k^2 + M^2}{M^2}\right) \right];$$

$$B_0(k, 0, 0) = -i\Delta + 2i\pi^2 - i\pi^2 \ln(k^2);$$

$$B_0(k, m, m) \big|_{k^2 = -m^2} = -i\Delta - i\pi^2 \left[\frac{\pi}{\sqrt{3}} - 2 + \ln(m^2) \right].$$

For large positive $s = -k^2$:

$$B_0(k, m, M) \big|_{k^2 \to -\infty} = - i\Delta - \pi^3 \left(1 - 2\epsilon - 2\epsilon^2 + 2\delta^2\right)$$

$$+ i\pi^2 \left[2 - \ln(s) - \delta \ln\left(\frac{M^2}{m^2}\right) \right.$$

$$\left. - (\epsilon + \epsilon^2 - \delta^2) \ln\left(\epsilon^2 - \delta^2\right) + 2\epsilon - \epsilon^2 - \delta^2 \right] + \mathcal{O}(s^{-3})$$

with

$$\epsilon = \frac{M^2 + m^2}{2s} \quad \text{and} \quad \delta = \frac{M^2 - m^2}{2s}.$$

To complete the review of the one-loop two-point function here are a few more equations. Define (remember F_1 above):

$$A(m) = im^2 \Delta + i\pi^2 m^2 \ln(m^2) - i\pi^2 m^2 = \int d_n p \frac{1}{p^2 + m^2};$$

$$k_\mu B_1(k, m, M) = \int d_n p \frac{p_\mu}{(p^2 + m^2)((p+k)^2 + M^2)};$$

$$k_\mu k_\nu B_{21}(k, m, M) + \delta_{\mu\nu} B_{22}(k, m, M) =$$
$$\int d_n p \frac{p_\mu p_\nu}{(p^2 + m^2)((p+k)^2 + M^2)}.$$

The functions B_1, B_{21} and B_{22} can be expressed in terms of the functions A and B_0. We will show how that works for B_1. Multiply the equation involving B_1 by k_μ. Then write

$$pk = \frac{1}{2} \left((p+k)^2 + M^2 - (p^2 + m^2) + m^2 - M^2 - k^2 \right).$$

This gives straightaway the equation shown below. To obtain equations for B_{12} and B_{22} one multiplies the equation involving them with k_μ or $\delta_{\mu\nu}$. The n arising from $\delta_{\mu\mu} = n$ must be treated with care, as shown earlier.

$$B_1(k, m, M) =$$
$$\frac{1}{2k^2} \left[A(m) - A(M) + (m^2 - M^2 - k^2) B_0(k, m, M) \right];$$

$$B_{21}(k, m, M) = \frac{i\pi^2}{18k^2} \left(3m^2 + 3M^2 + k^2 \right)$$
$$+ \frac{1}{3k^4} \left(m^2 - M^2 - k^2 \right) A(m)$$
$$+ \frac{1}{3k^4} \left(M^2 - m^2 + 2k^2 \right) A(M)$$
$$+ \frac{1}{3} \left(1 + \frac{2M^2 - m^2}{k^2} + \frac{(M^2 - m^2)^2}{k^4} \right) B_0(k, m, M);$$

$$B_{22}(k, m, M) = - \frac{i\pi^2}{18} \left(3m^2 + 3M^2 + k^2 \right)$$
$$+ \frac{1}{12k^2} \left(M^2 - m^2 + k^2 \right) A(m)$$

$$+ \frac{1}{12k^2} \left(m^2 - M^2 + k^2\right) A(M)$$

$$+ \frac{1}{6} \left(-m^2 - M^2 - \frac{k^2}{2} + \frac{(M^2 - m^2)^2}{2k^2}\right) B_0(k, m, M).$$

The infinite parts of these functions are $(\Delta = 2\pi^2/(n-4) + c)$:

$$A(m) : im^2\Delta, \qquad B_0 : -i\Delta, \qquad B_1 : \frac{i}{2}\Delta,$$

$$B_{21} : -\frac{i}{3}\Delta, \qquad B_{22} : \frac{i}{12}(k^2 + 3m^2 + 3M^2)\Delta.$$

In the literature these same functions are often used as above but divided by a factor $i\pi^2$, e.g. $A(\text{lit}) = A(\text{above}) / i\pi^2$.

Appendix D
Summary. Combinatorial Factors

D.1 Summary

The decay rate of a particle of momentum p into particles with momenta k_1, k_2, ... is given by:

$$\Gamma = \int \frac{V d_3 k_1}{(2\pi)^3} \int \frac{V d_3 k_2}{(2\pi)^3} \cdots$$

$$P \frac{V}{(2\pi)^4} |\mathcal{M}|^2 \delta_4(p - k_1 - k_2 - \ldots).$$

The reduced matrix element $\mathcal{M}$ is equal to the S-matrix element apart from the δ-function for overall energy–momentum conservation. The permutational factor P relates to identical particles in the final state: it contains a factor $1/n!$ for every set of n identical particles (bosons). For example, if there are three identical particles in the final state it is $1/3!$, and if there are five particles, of which two identical of one type and three identical of another type it is $1/2!3!$. Concerning Lorentz invariance, the reduced matrix element is of the form:

$$\mathcal{M} = \frac{X}{\sqrt{2V p_0 \, 2V k_{10} \, 2V k_{20} \ldots}}$$

where X is a Lorentz invariant expression. After squaring, a factor such as $1/2V k_0$ may be combined with $V d_3 k$ to give the Lorentz invariant measure $d_3 k / 2k_0$. Of course, the energy $k_0 = \sqrt{\vec{k}^2 + m^2}$. Formally, one has

$$\int \frac{d_3 k}{2k_0} = \int d_4 k \, \theta(k_0) \, \delta(k^2 + m^2).$$

On the right hand side k_0 is now integration variable, $d_4 k = d_3 k \, d k_0$, and $k^2 = \vec{k}^2 - k_0^2$ is the four-dimensional dot-product of

243

k with itself. Using polar coordinates:

$$\int d_3k = \int_0^\infty k_\ell^2 dk_\ell \int_{-1}^1 d\cos(\theta) \int_0^{2\pi} d\varphi,$$

where k_ℓ is the length of $\vec{k}$. A useful relation in this context is $k_\ell dk_\ell = k_0 dk_0$.

Expressing everything in MeV the decay rate Γ is expressed in MeV. The lifetime $\tau = 1/\Gamma$, and to get the result in seconds multiply by $\hbar = 6.582122 \times 10^{-22}$ MeV sec.

The total cross section for the collision of two particles of momenta p_1 and p_2 to particles with momenta k_1, k_2, ... is:

$$\sigma_{\text{tot}} = \int \frac{V d_3 k_1}{(2\pi)^3} \int \frac{V d_3 k_2}{(2\pi)^3} \cdots$$

$$F \cdot P \frac{V}{(2\pi)^4} |\mathcal{M}|^2 \delta_4(p_1 + p_2 - k_1 - k_2 - \ldots).$$

The main difference with the expression for a decay rate is the flux factor F. This factor is:

$$F = V \sqrt{\frac{p_{10}^2 p_{20}^2}{(p_1 p_2)^2 - m_1^2 m_2^2}}.$$

where m_1 and m_2 are the masses of the particles with momenta p_1 and p_2 respectively. In the laboratory system where the particle with momentum p_2 is at rest this reduces to:

$$F = \frac{V p_{10}}{|\vec{p}_1|} \qquad \text{(Lab system)}.$$

Expressing everything in MeV the total cross section is expressed in MeV^{-2}. To have the result in cm^2 multiply by $(\hbar c)^2 = (1.97327 \times 10^{-11}$ MeV cm$)^2$.

The matrix-element $\mathcal{M}$ may be computed using the Feynman rules for the theory considered. These rules involve vertices depending on the theory considered. For the Standard Model they are given in another appendix, and they include a δ-function for energy–momentum conservation not explicitly shown in that appendix. There is an integral over the momentum of every propagator; integrating over momenta in so far as the elimination of these vertex δ-functions permit leaves a final expression containing integration over closed loop momenta only. In addition there are propagators and external line factors, shown separately in a section below. The propagators shown in that section include the factor given under (ii). The factors are:

(i) $(2\pi)^4 i$ for every vertex;
(ii) $1/(2\pi)^4 i$ for every propagator;
(iii) $(-1)^\ell$ for diagrams with fermion lines;
(iv) A combinatorial factor, discussed below;
(v) A factor -1 for every fermion and FP ghost loop.

A factor -1 for every incoming anti-fermion is included in the prescription for the external line factors. That factor is, through unitarity, closely related to the -1 for every fermion loop.

The sign mentioned under (iii) has been discussed at length in the specification of the Feynman rules for quantum electrodynamics; it provides for proper anti-symmetrization with respect to fermions. Generally two diagrams differing by exchange of two fermion lines, internal or external, differ by a sign. However, the minus sign related to exchange of an incoming and an outgoing fermion is included in the prescription of a minus sign for an incoming anti-fermion. Diagrams differing by addition of a boson line have the same ℓ.

D.2 External Lines, Spin Sums, Propagators

Convention: the momentum for an incoming or outgoing particle is always directed such that its energy is positive, i.e., inwards for incoming particles, outwards for outgoing particles.

Scalar particles

External lines: $\qquad\qquad 1/\sqrt{2Vk_0}$

Propagator: $\qquad\qquad \dfrac{1}{(2\pi)^4 i}\dfrac{1}{k^2 + m^2 - i\epsilon}$

In shadowed region: $\qquad \dfrac{-1}{(2\pi)^4 i}\dfrac{1}{k^2 + m^2 + i\epsilon}$

Cut propagator: $\qquad\quad \dfrac{1}{(2\pi)^3}\,\theta(k_0)\delta(k^2 + m^2)$

Spin 1/2 particles

Incoming particle: $\qquad u^a(k)/\sqrt{V}$, $a = 1,2$

Incoming anti-particle: $-\bar{u}^a(k)/\sqrt{V}$, $a = 3,4$

Outgoing particle: $\qquad \bar{u}^a(k)/\sqrt{V}$, $a = 1,2$

Outgoing anti-particle: $\qquad u^a(k)/\sqrt{V}$, $a = 3,4$

Spinsum particle: $\displaystyle\sum_{j=1}^{2} u^j(k)\bar{u}^j(k) = \frac{1}{2k_0}(-i\gamma k + m)$

Spinsum anti-particle: $\displaystyle\sum_{j=3}^{4} u^j(k)\bar{u}^j(k) = \frac{1}{2k_0}(-i\gamma k - m)$

Propagator: $\dfrac{1}{(2\pi)^4 i}\dfrac{-i\gamma k + m}{k^2 + m^2 - i\epsilon}$

In shadowed region: $\dfrac{-1}{(2\pi)^4 i}\dfrac{-i\gamma k + m}{k^2 + m^2 + i\epsilon}$

Cut propagator: $\dfrac{1}{(2\pi)^3}\,(-i\gamma k + m)\,\theta(-k_0)\delta(k^2 + m^2)$

$\dfrac{1}{(2\pi)^3}\,(-i\gamma k + m)\,\theta(k_0)\delta(k^2 + m^2)$

Vector particles

incoming particle: $\qquad e_\mu(k)/\sqrt{2Vk_0}$

outging particle: $\qquad \bar{e}_\mu(k)/\sqrt{2Vk_0}$

Normalization: $\qquad e_\mu \bar{e}_\mu = 1$

$\qquad\qquad\qquad \bar{e}_\mu = (e_\mu)^*$ for $\mu = 1\text{-}3$, and $\bar{e}_4 = -(e_4)^*$

Spinsum: $\displaystyle\sum_{j=1}^{3} e_\mu^j \bar{e}_\nu^j = \delta_{\mu\nu} + \frac{k_\mu k_\nu}{m^2}$

Propagator: $\dfrac{1}{(2\pi)^4 i}\dfrac{\delta_{\mu\nu} + k_\mu k_\nu/m^2}{k^2 + m^2 - i\epsilon}$

In shadowed region: $\dfrac{-1}{(2\pi)^4 i}\dfrac{\delta_{\mu\nu} + k_\mu k_\nu/m^2}{k^2 + m^2 + i\epsilon}$

Cut propagator: $\dfrac{1}{(2\pi)^3}\left(\delta_{\mu\nu} + k_\mu k_\nu/m^2\right)\theta(k_0)\delta(k^2 + m^2)$

For the massless photon use the above, omitting terms $k_\mu k_\nu/m^2$ and setting $m = 0$.

D.3 Combinatorial Factors

The combinatorial factors are sometimes quite complicated factors to obtain. They can be found by brute force, i.e., directly using the procedure suggested by the time-ordered product of interaction Lagrangians. That is what we will show here. From this one may derive simpler rules, allowing one to establish these factors on the basis of symmetry considerations. We refer to the literature for that procedure (G. Goldberg, Phys. Rev. 15 (1985) 3331).

We will explain things by way of examples. No derivatives are considered; we only note that lines with derivatives are to be distinguished from lines without derivatives. For lines with arrows the direction of the arrow is relevant.

The example to be considered involves two vertices, a three-point and a four-point vertex. All fields are identical. Given the interactions

$$g\phi^3 \quad \text{and} \quad g\phi^4$$

one has the vertex factors $(2\pi)^4 i\,\alpha$ and $(2\pi)^4 i\,\beta$, where $\alpha = g\,3!$ and $\beta = g\,4!$. Thus remember, the factorials are normally part of the vertex definition. They are later divided out, in the procedure below.

Consider the lowest order self-energy diagram, as shown.

Draw two points, x_1 and x_2, and draw in each of these two points the three-point vertex.

Count the number of ways that a line can be connected such that the same topology results. The external line 1 can be connected in six ways, after that line 2 in three ways. The remaining

248   *Summary. Combinatorial Factors*

lines can be connected in two different ways. Altogether $6 \times 3 \times 2$ combinations. We must divide by the factorials from the vertices (two times 3!) and the number of permutations of points having identical vertices, here 2!. The net result is:

$$\frac{6 \times 3 \times 2}{3!\,3!\,2!} = \frac{1}{2}.$$

That is the combinatorial factor for this self-energy diagram. For comparison: in quantum electrodynamics, for a photon self-energy or an electron self-energy diagram the factor is $2/2! = 1$.

Next consider a two loop diagram, the first shown in the figure. Draw three points and the lines attached, see the next figure. Line 1 can be connected in six ways, line 2 in four ways. The result up to that point is shown in the third figure. After this there are $6 \times 3 \times 2$ ways to connect the rest. The whole must be divided by the vertex factors, 3!, 3! and 4!, and by the permutation factor for identical vertices, 2!. The final result is:

$$\frac{6 \times 4 \times 6 \times 3 \times 2}{3!\,3!\,4!\,2!} = \frac{1}{2}.$$

The second two loop diagram shown in the above figure is slightly more complicated. Drawing the vertical line last the result is:

$$\frac{12 \times 9 \times 6 \times 3 \times 4 \times 2}{3!\,3!\,3!\,3!\,4!} = \frac{1}{2}.$$

The same topology, for quantum electrodynamics, with two external photon lines and a photon line for the vertical, gives as always just 1.

We leave it to the reader to establish that the factor for tadpole, like diagrams as shown is 1/2.

Finally, for the case of two identical sources coupled to a scalar line, the factor is 1/2!.

Appendix E
Standard Model

E.1 Lagrangian

The Lagrangian of the Standard Model is quite voluminous, and it is already a problem by itself to get all the Feynman rules correct. We will give it a try in this appendix. This section gives the explicit form of the Lagrangian. We assume the simplest Higgs sector. The gauge chosen is the Feynman–'t Hooft gauge. In this gauge the numerator of the vector boson propagators is of the form $\delta_{\mu\nu}$ with respect to the Lorentz indices. There are ghost fields, Higgs ghosts and Faddeev–Popov ghosts. The ghost fields must be included for internal lines, but they should not occur as external lines. They do not correspond to physical particles, but they occur in the diagrams to correct violations of unitarity that would otherwise arise due to the form of the vector boson propagators chosen here. The proof of that fact is really the central part of gauge field theory.

The Lagrangian including the gauge breaking terms can conveniently be subdivided in a number of pieces:

$$\mathcal{L}_{\text{total}} = \mathcal{L}_{\text{c}} + \mathcal{L}_{\text{w}} + \mathcal{L}_{\text{f}} + \mathcal{L}_{\text{fH}} + \mathcal{L}_{\text{fc}} + \mathcal{L}_{\text{FPc}} + \mathcal{L}_{\text{FPw}} \, .$$

These pieces are

- $\mathcal{L}_{\text{c}}$: colour Lagrangian, describing the gluons of quantum chromodynamics and their mutual interactions;
- $\mathcal{L}_{\text{w}}$: weak Lagrangian, describing the vector bosons and their interactions including interactions with the Higgs system;
- $\mathcal{L}_{\text{f}}$: fermion Lagrangian, describing the interactions of the fermions with the weak vector bosons;
- $\mathcal{L}_{\text{fH}}$: fermion–Higgs Lagrangian, describing the interactions of the fermions with the Higgs system;

249

- $\mathcal{L}_{\rm fc}$: fermion–colour Lagrangian, describing the interactions of the fermions with the gluons;
- $\mathcal{L}_{\rm FPc}$: the Faddeev–Popov ghost Lagrangian of quantum chromodynamics;
- $\mathcal{L}_{\rm FPw}$: the Faddeev–Popov ghost Lagrangian of the weak interactions.

A feature of the present day situation must be mentioned here. In the old days, doing quantum electrodynamics, the important feature was the bare simplicity of the Feynman rules and the small number of particles. The electron and/or muon and the photon were the main players. One vertex, the fermion–photon vertex was all that was needed. As a consequence the number of graphs remained relatively small, in fact usually only one at the one loop level. This fact has made the calculation of as much as four loop effects possible. These very advanced calculations require the use of mechanical procedures, i.e., algebraic computer programs.

To some extent this same simplicity applies to quantum chromodynamics. But in the weak interactions the situation is vastly different. There are now many particles to be considered, and the number of vertices is huge. As a consequence the number of one loop graphs is often already so large that mechanical procedures are needed to process them. These mechanical procedures are very different in nature from those used in quantum electrodynamics. In weak interactions the difficulty is in handling the multitude of vertices and particles.

To establish notation, here are the fields corresponding to the known particles at this moment and the various ghosts.

- A_μ: the photon;
- W_μ^+, W_μ^-, Z_μ^0: the charged and neutral vector bosons of weak interactions;
- ϕ^+, ϕ^-, ϕ^0: the charged and neutral Higgs ghosts;
- H: the physical Higgs particle;
- e^α, ν^α, $\alpha = 1, 2, 3$: three lepton generations;
- d_j^α, u_j^α, $\alpha = 1, 2, 3$, $j = 1, 2, 3$: three quark generations of three colours each.
- Y: the FP ghost associated with the photon;
- X^+, X^- and X^0: the FP ghosts associated with the three vector bosons of weak interactions;

- g_μ^a, $a = 1 \ldots 8$: the eight gluons;
- G^a, $a = 1 \ldots 8$: the eight FP ghosts associated with the gluons.

The FP (= Faddeev–Popov) ghost fields are larger in number than one might think at first. To begin with, remember that they are fictitious particles with fictitious rules. The minus sign for a closed FP loop is just a prescription. Further, the X^- field has nothing to do with the X^+ field, for example it is not true that the X^+ field contains a creation operator for an X^- particle. The way this must be read is this: the X^- field contains the absorption operator for an X^- and the creation operator for an anti-X^-. The anti-X^- field will be denoted by $\bar{X}^-$ and contains the absorption operator for an anti-X^- and the creation operator for an X^-. So, on the one loop level, considering Z^0 self-energy diagrams there will be typically two X-graphs: one with an X^- and one with an X^+ circulating. By contrast, there is only one graph with a circulating charged vector boson. If we had used real vector bosons W^1 and W^2 there would have been two W graphs, and that indeed corresponds more closely to the situation in the FP Lagrangian. In fact, the rule is that corresponding to a real vector field there is a complex FP field. In the case at hand that applies to the FP ghosts associated with the photon and gluon fields. Thus there are $\bar{Y}$ and $\bar{G}^\alpha$ fields.

The parameters occurring in this Lagrangian will be discussed below.

$$\mathcal{L}_w =$$

$$-\partial_\nu W_\mu^+ \partial_\nu W_\mu^- - M^2 W_\mu^+ W_\mu^- \qquad -\tfrac{1}{2}\partial_\nu Z_\mu^0 \partial_\nu Z_\mu^0 - \tfrac{1}{2c_w^2} M^2 Z_\mu^0 Z_\mu^0$$

$$-\tfrac{1}{2}\partial_\mu A_\nu \partial_\mu A_\nu \qquad\qquad -\tfrac{1}{2}\partial_\mu H \partial_\mu H - \tfrac{1}{2}m_h^2 H^2$$

$$-\partial_\mu \phi^+ \partial_\mu \phi^- - M^2 \phi^+ \phi^- \qquad -\tfrac{1}{2}\partial_\mu \phi^0 \partial_\mu \phi^0 - \tfrac{1}{2c_w^2} M \phi^0 \phi^0$$

$$-\beta_h \left[\frac{2M^2}{g^2} + \frac{2M}{g} H + \tfrac{1}{2}(H^2 + \phi^0 \phi^0 + 2\phi^+ \phi^-) \right] + \frac{2M^4}{g^2}\alpha_h$$

$$-igc_w [\partial_\nu Z_\mu^0 \left(W_\mu^+ W_\nu^- - W_\nu^+ W_\mu^- \right) - Z_\nu^0 \left(W_\mu^+ \partial_\nu W_\mu^- - W_\mu^- \partial_\nu W_\mu^+ \right)$$

$$+ Z_\mu^0 \left(W_\nu^+ \partial_\nu W_\mu^- - W_\nu^- \partial_\nu W_\mu^+ \right)]$$

$$-igs_w[\partial_\nu A_\mu \left(W_\mu^+ W_\nu^- - W_\nu^+ W_\mu^-\right) - A_\nu \left(W_\mu^+ \partial_\nu W_\mu^- - W_\mu^- \partial_\nu W_\mu^+\right)$$
$$+A_\mu \left(W_\nu^+ \partial_\nu W_\mu^- - W_\nu^- \partial_\nu W_\mu^+\right)]$$

$$-\tfrac{1}{2}g^2 W_\mu^+ W_\mu^- W_\nu^+ W_\nu^- + \tfrac{1}{2}g^2 W_\mu^+ W_\nu^- W_\mu^+ W_\nu^-$$

$$+g^2 c_w^2 \left(Z_\mu^0 W_\mu^+ Z_\nu^0 W_\nu^- - Z_\mu^0 Z_\mu^0 W_\nu^+ W_\nu^-\right)$$

$$+g^2 s_w^2 \left(A_\mu W_\mu^+ A_\nu W_\nu^- - A_\mu A_\mu W_\nu^+ W_\nu^-\right)$$

$$+g^2 s_w c_w [A_\mu Z_\nu^0 (W_\mu^+ W_\nu^- + W_\nu^+ W_\mu^-) - 2A_\mu Z_\mu^0 W_\nu^+ W_\nu^-]$$

$$-g\alpha_h M[H^3 + H\phi^0 \phi^0 + 2H\phi^+ \phi^-]$$

$$-\tfrac{1}{8}g^2 \alpha_h [H^4 + (\phi^0)^4 + 4(\phi^+ \phi^-)^2 + 4(\phi^0)^2 \phi^+ \phi^-$$
$$+4H^2 \phi^+ \phi^- + 2(\phi^0)^2 H^2]$$

$$-gMW_\mu^+ W_\mu^- H - \tfrac{1}{2}g\tfrac{M}{c_w^2}Z_\mu^0 Z_\mu^0 H$$

$$-\tfrac{1}{2}ig[W_\mu^+ (\phi^0 \partial_\mu \phi^- - \phi^- \partial_\mu \phi^0) - W_\mu^- (\phi^0 \partial_\mu \phi^+ - \phi^+ \partial_\mu \phi^0)]$$

$$+\tfrac{1}{2}g[W_\mu^+ (H\partial_\mu \phi^- - \phi^- \partial_\mu H) + W_\mu^- (H\partial_\mu \phi^+ - \phi^+ \partial_\mu H)]$$

$$+\tfrac{1}{2}g\tfrac{1}{c_w}Z_\mu^0 (H\partial_\mu \phi^0 - \phi^0 \partial_\mu H)$$

$$-ig\tfrac{s_w^2}{c_w}MZ_\mu^0 (W_\mu^+ \phi^- - W_\mu^- \phi^+) \quad +igs_w MA_\mu (W_\mu^+ \phi^- - W_\mu^- \phi^+)$$

$$-ig\tfrac{1-2c_w^2}{2c_w}Z_\mu^0 (\phi^+ \partial_\mu \phi^- - \phi^- \partial_\mu \phi^+) + igs_w A_\mu (\phi^+ \partial_\mu \phi^- - \phi^- \partial_\mu \phi^+)$$

$$-\tfrac{1}{4}g^2 W_\mu^+ W_\mu^- [H^2 + (\phi^0)^2 + 2\phi^+ \phi^-]$$

$$-\tfrac{1}{8}g^2 \tfrac{1}{c_w^2}Z_\mu^0 Z_\mu^0 [H^2 + (\phi^0)^2 + 2(2s_w^2 - 1)^2 \phi^+ \phi^-]$$

$$-\tfrac{1}{2}g^2 \tfrac{s_w^2}{c_w}Z_\mu^0 \phi^0 (W_\mu^+ \phi^- + W_\mu^- \phi^+) - \tfrac{1}{2}ig^2 \tfrac{s_w^2}{c_w}Z_\mu^0 H(W_\mu^+ \phi^- - W_\mu^- \phi^+)$$

$$+\tfrac{1}{2}g^2 s_w A_\mu \phi^0 (W_\mu^+ \phi^- + W_\mu^- \phi^+)$$

$$+\tfrac{1}{2}ig^2 s_w A_\mu H(W_\mu^+ \phi^- - W_\mu^- \phi^+)$$

$$-g^2 \tfrac{s_w}{c_w}(2c_w^2 - 1)Z_\mu^0 A_\mu \phi^+ \phi^- \quad -g^2 s_w^2 A_\mu A_\mu \phi^+ \phi^-.$$

The quantities appearing in this Lagrangian are given below. The numerical values are from the Particle Properties Data table 1992. Numbers depend on values assumed for the top quark and

Higgs masses, and there is some ambiguity of interpretation if it comes to including radiative corrections, because to some extent these can be pushed around. The experimentally measured quantities include of course all radiative corrections, and must not be confused with the parameters in the Lagrangian. Consider the numbers given below as indicative, up to the level of, say, 0.5 % .

- The sine and cosine of the weak mixing angle, s_w and c_w. Numerically $s_w^2 = 0.2337\text{--}0.2310$.
- The coupling constant g. Its numerical value is linked to the value of the fine-structure constant $\alpha_{\text{em}} = 1/137.036$ by means of the relation $\alpha_{\text{w}} \equiv g^2/4\pi = \alpha/s_w^2$. This gives $\alpha_{\text{w}} = 1/31.8$.
- The vector boson mass $M = 80.22 \pm 0.26$ GeV. Being an unstable particle the definition of its mass is ambiguous on the level of 0.1%.
- Experimentally the neutral vector boson mass $M_0 = 91.173 \pm 0.020$ GeV. That includes of course all radiative corrections. From the point of view of parameters in the Lagrangian, with the simplest Higgs system used here, the neutral vector boson mass M_0 is not a free parameter, and equals the mass of the charged vector boson M divided by c_w: $M_0 = M/c_w$.
- The Higgs mass m_h. Only a lower limit is known: $m_h > 50$ GeV.
- The tadpole constant β_h. It is zero in lowest order, and must be adjusted such that the vacuum expectation value of the Higgs field H remains zero.
- The Higgs scattering parameter α_h. This is not an independent parameter, it is equal to $m_h^2/4M^2$. If this parameter exceeds $\sqrt{1/\alpha_w} \approx 6$ perturbation theory for the Higgs sector breaks down.

The fermion Lagrangian describes the interactions of three lepton and three quark generations. Below, zero neutrino mass is assumed. The generations are labelled using the indices λ and κ, and for example e^2 is the muon, and ν^3 is the τ-neutrino. Likewise u^3 is the top quark. Colour is indexed by means of the index j, taking the values 1–3. The various parameters, including the unitary matrix C (CKM matrix) will be discussed below.

$$\mathcal{L}_f =$$

$$-\bar{e}^\lambda(\gamma\partial + m_e^\lambda)e^\lambda - \bar{\nu}^\lambda\gamma\partial\nu^\lambda - \bar{u}_j^\lambda(\gamma\partial + m_u^\lambda)u_j^\lambda - \bar{d}_j^\lambda(\gamma\partial + m_d^\lambda)d_j^\lambda$$

$$+igs_w A_\mu[-(\bar{e}^\lambda\gamma^\mu e^\lambda) + \tfrac{2}{3}(\bar{u}_j^\lambda\gamma^\mu u_j^\lambda) - \tfrac{1}{3}(\bar{d}_j^\lambda\gamma^\mu d_j^\lambda)]$$

$$+\tfrac{ig}{4c_w}Z_\mu^0[\,(\bar{\nu}^\lambda\gamma^\mu(1+\gamma^5)\nu^\lambda) + (\bar{e}^\lambda\gamma^\mu(4s_w^2 - 1 - \gamma^5)e^\lambda)$$

$$+(\bar{d}_j^\lambda\gamma^\mu(\tfrac{4}{3}s_w^2 - 1 - \gamma^5)d_j^\lambda) + (\bar{u}_j^\lambda\gamma^\mu(1 - \tfrac{8}{3}s_w^2 + \gamma^5)u_j^\lambda)]$$

$$+\tfrac{ig}{2\sqrt{2}}W_\mu^+[\,(\bar{\nu}^\lambda\gamma^\mu(1+\gamma^5)e^\lambda) + (\bar{u}_j^\lambda\gamma^\mu(1+\gamma^5)C_{\lambda\kappa}d_j^\kappa)]$$

$$+\tfrac{ig}{2\sqrt{2}}W_\mu^-[\,(\bar{e}^\lambda\gamma^\mu(1+\gamma^5)\nu^\lambda) + (\bar{d}_j^\kappa C_{\kappa\lambda}^\dagger\gamma^\mu(1+\gamma^5)u_j^\lambda)]$$

The Fermion–Higgs Lagrangian contains the interactions of the Higgs ghosts with the fermions, and the essentially untested interactions with the Higgs particle. All these interactions are proportional to the ratio of the fermion mass and the vector boson mass.

$$\mathcal{L}_{fH} =$$

$$\tfrac{ig}{2\sqrt{2}}\tfrac{m_e^\lambda}{M}[-\phi^+(\bar{\nu}^\lambda(1-\gamma^5)e^\lambda) + \phi^-(\bar{e}^\lambda(1+\gamma^5)\nu^\lambda)]$$

$$-\tfrac{g}{2}\tfrac{m_e^\lambda}{M}[H(\bar{e}^\lambda e^\lambda) + i\phi^0(\bar{e}^\lambda\gamma^5 e^\lambda)]$$

$$+\tfrac{ig}{2M\sqrt{2}}\phi^+\left[-m_d^\kappa(\bar{u}_j^\lambda C_{\lambda\kappa}(1-\gamma^5)d_j^\kappa) + m_u^\lambda(\bar{u}_j^\lambda C_{\lambda\kappa}(1+\gamma^5)d_j^\kappa)\right]$$

$$+\tfrac{ig}{2M\sqrt{2}}\phi^-\left[m_d^\lambda(\bar{d}_j^\lambda C_{\lambda\kappa}^\dagger(1+\gamma^5)u_j^\kappa) - m_u^\kappa(\bar{d}_j^\lambda C_{\lambda\kappa}^\dagger(1-\gamma^5)u_j^\kappa)\right]$$

$$-\tfrac{g}{2}\tfrac{m_u^\lambda}{M}H(\bar{u}_j^\lambda u_j^\lambda) - \tfrac{g}{2}\tfrac{m_d^\lambda}{M}H(\bar{d}_j^\lambda d_j^\lambda) + \tfrac{ig}{2}\tfrac{m_u^\lambda}{M}\phi^0(\bar{u}_j^\lambda\gamma^5 u_j^\lambda)$$

$$-\tfrac{ig}{2}\tfrac{m_d^\lambda}{M}\phi^0(\bar{d}_j^\lambda\gamma^5 d_j^\lambda).$$

The various quantities in the fermion Lagrangian are:

- The lepton masses: $m_e = 0.511$ MeV, $m_\mu = 105.658$ MeV and $m_\tau = 1784$ MeV.
- The quark masses: $m_d = 5\text{–}15$ MeV, $m_u = 2\text{–}8$ MeV, $m_s = 100\text{–}300$ MeV, $m_c = 1.3\text{–}1.7$ GeV, $m_b = 4.7\text{–}5.3$ GeV and $m_t > 91$ GeV.
- The Cabibbo–Kobayashi–Maskawa matrix, specified below.

The most general form of the quark mixing matrix may be transformed into a unitary matrix C, using the symmetries of the Lagrangian. This matrix can be parametrized in various ways; we follow the conventions of the Data booklet. In analysing data, and in the Feynman rules one uses the notation shown here:

$$C = \begin{bmatrix} V_{ud} & V_{us} & V_{ub} \\ V_{cd} & V_{cs} & V_{cb} \\ V_{td} & V_{ts} & V_{tb} \end{bmatrix}$$

Experimentally, the ranges for the various parameters are:

$$C = \begin{bmatrix} 0.9747\text{–}0.9759 & 0.218\text{–}0.224 & 0.002\text{–}0.007 \\ 0.218\text{–}0.224 & 0.9735\text{–}0.9751 & 0.032\text{–}0.054 \\ 0.003\text{–}0.018 & 0.030\text{–}0.054 & 0.9985\text{–}0.9995 \end{bmatrix}$$

In terms of three angles, θ_{12}, θ_{13} and θ_{23}, all in the first quadrant, and a phase δ, in the range 0–2π, one has:

$$C = \begin{bmatrix} c_{12}c_{13} & s_{12}c_{13} & s_{13}e^{-i\delta} \\ -s_{12}c_{23} - c_{12}s_{23}s_{13}e^{i\delta} & c_{12}c_{23} - s_{12}s_{23}s_{13}e^{i\delta} & s_{23}c_{13} \\ s_{12}s_{23} - c_{12}c_{23}s_{13}e^{i\delta} & -c_{12}s_{23} - s_{12}c_{23}s_{13}e^{i\delta} & c_{23}c_{13} \end{bmatrix}$$

As the reader surely guessed, $s_{12} = \sin\theta_{12}$ etc. The phase angle δ is usually denoted as δ_{13}. The experimental ranges for the angles are at this time:

$$s_{12} = 0.218\text{–}0.224 \quad s_{23} = 0.032\text{–}0.054 \quad s_{13} = 0.002\text{–}0.007 .$$

The quantity δ_{13} is essentially unknown, except that it is non-zero if indeed CP-violation as observed in K-decays is due to this phase. The angle θ_{12} is very close to what used to be the Cabibbo angle, at a time when there was no third generation on the horizon. We repeat that the matrix C is unitary, i.e., $C^\dagger = (\widetilde{C})^* = C^{-1}$.

The weak Faddeev–Popov ghost Lagrangian is quite complicated, but contains no references to fermions. One must be thankful for little things.

$$\mathcal{L}_{\text{FPw}} =$$

$$\bar{X}^+(\partial^2 - M^2)X^+ + \bar{X}^-(\partial^2 - M^2)X^- + \bar{X}^0(\partial^2 - \tfrac{M^2}{c_w^2})X^0 + \bar{Y}\partial^2 Y$$

$$+ igc_w W_\mu^+(\partial_\mu \bar{X}^0 X^- - \partial_\mu \bar{X}^+ X^0) + igs_w W_\mu^+(\partial_\mu \bar{Y} X^- - \partial_\mu \bar{X}^+ Y)$$

$$+ igc_w W_\mu^-(\partial_\mu \bar{X}^- X^0 - \partial_\mu \bar{X}^0 X^+) + igs_w W_\mu^-(\partial_\mu \bar{X}^- Y - \partial_\mu \bar{Y} X^+)$$

$$+ igc_w Z_\mu^0(\partial_\mu \bar{X}^+ X^+ - \partial_\mu \bar{X}^- X^-)$$

$$+ igs_w A_\mu(\partial_\mu \bar{X}^+ X^+ - \partial_\mu \bar{X}^- X^-)$$

$$- \tfrac{1}{2}gM \left[\bar{X}^+ X^+ H + \bar{X}^- X^- H + \tfrac{1}{c_w^2}\bar{X}^0 X^0 H \right]$$

$$+ \tfrac{1-2c_w^2}{2c_w}igM \left[\bar{X}^+ X^0 \phi^+ - \bar{X}^- X^0 \phi^- \right]$$

$$+ \tfrac{1}{2c_w}igM \left[\bar{X}^0 X^- \phi^+ - \bar{X}^0 X^+ \phi^- \right]$$

$$+ igMs_w \left[\bar{X}^- Y \phi^- - \bar{X}^+ Y \phi^+ \right] + \tfrac{1}{2}igM \left[\bar{X}^+ X^+ \phi^0 - \bar{X}^- X^- \phi^0 \right]$$

Quantum chromodynamics has become quite a domain of specialists. At low energies perturbation theory is not valid, and even at high energies, where perturbative QCD is supposedly valid the situation remains difficult. This is due to colour confinement, the large number of fundamental vector bosons (the gluons), and the fact that they are massless. Anyway, we just cite the Lagrangians for what they are worth. This requires a few preliminaries.

There are eight 3×3 hermitian traceless matrices λ^a, $a = 1 \ldots 8$. They are the straight generalization of the 2×2 Pauli spin matrices, in fact the first three are the Pauli spin matrices in a two dimensional subspace:

$$\lambda^1 = \begin{pmatrix} 0 & 1 & 0 \\ 1 & 0 & 0 \\ 0 & 0 & 0 \end{pmatrix} \qquad \lambda^4 = \begin{pmatrix} 0 & 0 & 1 \\ 0 & 0 & 0 \\ 1 & 0 & 0 \end{pmatrix} \qquad \lambda^6 = \begin{pmatrix} 0 & 0 & 0 \\ 0 & 0 & 1 \\ 0 & 1 & 0 \end{pmatrix}$$

$$\lambda^2 = \begin{pmatrix} 0 & -i & 0 \\ i & 0 & 0 \\ 0 & 0 & 0 \end{pmatrix} \qquad \lambda^5 = \begin{pmatrix} 0 & 0 & -i \\ 0 & 0 & 0 \\ i & 0 & 0 \end{pmatrix} \qquad \lambda^7 = \begin{pmatrix} 0 & 0 & 0 \\ 0 & 0 & -i \\ 0 & i & 0 \end{pmatrix}$$

$$\lambda^3 = \begin{pmatrix} 1 & 0 & 0 \\ 0 & -1 & 0 \\ 0 & 0 & 0 \end{pmatrix} \qquad \qquad \lambda^8 = \tfrac{1}{\sqrt{3}} \begin{pmatrix} 1 & 0 & 0 \\ 0 & 1 & 0 \\ 0 & 0 & -2 \end{pmatrix}$$

The commutation and anticommutation rules for these matrices are:

$$\left[\frac{-i}{2}\lambda^a, \frac{-i}{2}\lambda^b\right] = f^{abc}\frac{-i}{2}\lambda^c$$

$$\left\{\frac{-i}{2}\lambda^a, \frac{-i}{2}\lambda^b\right\} = -id^{abc}\frac{-i}{2}\lambda^c - \frac{1}{3}\delta_{ab}\lambda^0$$

where λ^0 is the 3×3 unit matrix. The coefficients f are antisymmetric in all three indices while the d are symmetric in all indices. They are:

f_{123}	f_{147}	f_{156}	f_{246}	f_{257}	f_{345}	f_{367}	f_{458}	f_{678}
1	$\frac{1}{2}$	$-\frac{1}{2}$	$\frac{1}{2}$	$\frac{1}{2}$	$\frac{1}{2}$	$-\frac{1}{2}$	$\frac{\sqrt{3}}{2}$	$\frac{\sqrt{3}}{2}$

d_{118}	d_{146}	d_{157}	d_{228}	d_{247}	d_{256}	d_{338}	d_{344}	d_{355}
$\frac{1}{\sqrt{3}}$	$\frac{1}{2}$	$\frac{1}{2}$	$\frac{1}{\sqrt{3}}$	$-\frac{1}{2}$	$\frac{1}{2}$	$\frac{1}{\sqrt{3}}$	$\frac{1}{2}$	$\frac{1}{2}$

d_{366}	d_{377}	d_{448}	d_{558}	d_{668}	d_{778}	d_{888}
$-\frac{1}{2}$	$-\frac{1}{2}$	$\frac{-1}{2\sqrt{3}}$	$\frac{-1}{2\sqrt{3}}$	$\frac{-1}{2\sqrt{3}}$	$\frac{-1}{2\sqrt{3}}$	$\frac{-1}{\sqrt{3}}$

The structure constants f satisfy a Jacobi identity:

$$f(g,a,b)f(g,c,d) + f(g,c,a)f(g,b,d) + f(g,b,c)f(g,a,d) = 0,$$

for any a, b, c and d and with g summed over from 1 to 8.

A useful equation for the trace of the product of two λ matrices:

$$\text{Tr}\left[\lambda^a\lambda^b\right] = 2\delta_{ab}.$$

The colour gluon, the colour fermion and the colour FP Lagrangian are:

$$\mathcal{L}_{\text{c}} = -\tfrac{1}{2}\partial_\nu g_\mu^a \partial_\nu g_\mu^a - g_s f^{abc}\partial_\mu g_\nu^a g_\mu^b g_\nu^c$$
$$\quad - \tfrac{1}{4}g_s^2 f^{abc}f^{ade} g_\mu^b g_\nu^c g_\mu^d g_\nu^e$$

$$\mathcal{L}_{\text{fc}} = \tfrac{1}{2}ig_s(\bar{q}_i^\sigma \gamma^\mu \lambda_{ij}^a q_j^\sigma)\, g_\mu^a$$

$$\mathcal{L}_{\text{FPc}} = \bar{G}^a\partial^2 G^a + g_s f^{abc}\partial_\mu \bar{G}^a G^b g_\mu^c$$

In here g_s is the (strong) coupling constant of QCD. The indices a, b and c take the values $1\ldots 8$ corresponding to the 8 gluons. The lower quark indices i and j take the values $1\ldots 3$, implying the three colours. The upper index σ, also to be summed over, designates the six quark flavours up, down, strange, charm, bottom and top. All these quarks have identical colour interactions.

E.2 Feynman Rules

The Lagrangian must now be translated into actual Feynman rules. Since there are many particles it is hardly sensible to try to invent a different linetype for every particle. We will do two things though: all ghosts will be denoted by dashed lines, and the Higgs will be a wavy line. The latter is traditionally reserved for the photon, but these days the Higgs is the outstanding one and deserves a special indication. We follow the convention of using a coil to denote gluons. For the rest we will simply show the particle symbol next to the line.

The arrows that occur in lines may be denoting fermion lines, or the flow of charge, or the flow of something that you could call FP-ghost number. The FP ghosts behave like fermions in the sense that a direction must be assigned and that a minus sign must be given to every closed FP-ghost loop.

In vertices all momenta are taken to be ingoing. An incoming W^+ is denoted by an incoming arrow, with the notation W associated with it. Similarly ϕ denotes an incoming ϕ^+, and the arrow points inwards to the vertex. In general, the arrow coming in is associated with annihilating a particle, and we just stated here that we take the W^+ and ϕ^+ to be particles (thus W^- and ϕ^- are anti-particles). Similarly, for FP ghosts or fermions an incoming arrow corresponds to the annihilation operator in an FP or fermion field.

Propagators

Gluon	$\dfrac{\delta_{\mu\nu}\delta_{ab}}{k^2 - i\epsilon}$	
Photon	$\dfrac{\delta_{\mu\nu}}{k^2 - i\epsilon}$	
W^+	$\dfrac{\delta_{\mu\nu}}{k^2 + M^2 - i\epsilon}$	
Z^0	$\dfrac{\delta_{\mu\nu}}{k^2 + M_0^2 - i\epsilon}$	

ϕ^+	$\dfrac{1}{k^2 + M^2 - i\epsilon}$	• - - -> $\overset{\phi}{\text{- - -}}$ •
ϕ^0	$\dfrac{1}{k^2 + M_0^2 - i\epsilon}$	• - - - - $\overset{\phi}{\text{- - -}}$ •
H	$\dfrac{1}{k^2 + m_h^2 - i\epsilon}$	$\sim\!\!\sim\!\!\sim\!\overset{H}{\sim\!\!\sim}$ •
Fermion	$\dfrac{-i\gamma k + m_f}{k^2 + m_f^2 - i\epsilon}$	• ————> e •
FP ghost	$\dfrac{1}{k^2 + M_x^2 - i\epsilon}$	• - - -> $\overset{X^+}{\text{- - -}}$ •

The FP masses are: M for X^+ and X^-, M_0 for X^0, 0 for Y and the G^a. Note that $M_0 = M/c_w$.

Vertices Involving Vector Bosons, Higgs

Those vertices that become non-perturbative for large Higgs mass m_h are marked in the figures by a little square. In the vertices of this part there are no fermions or FP ghosts, and the arrows indicate flow of charge. An arrow pointing inwards implies positive charge flowing into the vertex. All momenta ingoing.

AW^+W^-	$gs_w[\,\delta_{\mu\nu}(p_\sigma - q_\sigma)$ $+ \delta_{\nu\sigma}(q_\mu - k_\mu)$ $+ \delta_{\mu\sigma}(k_\nu - p_\nu)\,]$	
$Z^0W^+W^-$	$gc_w[\,\delta_{\mu\nu}(p_\sigma - q_\sigma)$ $+ \delta_{\nu\sigma}(q_\mu - k_\mu)$ $+ \delta_{\mu\sigma}(k_\nu - p_\nu)\,]$	

$AW^-\phi^+$

$$-igs_w M\delta_{\mu\nu}$$

$AW^+\phi^-$

$$igs_w M\delta_{\mu\nu}$$

$Z^0W^-\phi^+$

$$ig\frac{s_w^2}{c_w}M\delta_{\mu\nu}$$

$Z^0W^+\phi^-$

$$-ig\frac{s_w^2}{c_w}M\delta_{\mu\nu}$$

W^-W^+H

$$-gM\delta_{\mu\nu}$$

Z^0Z^0H

$$-gM\frac{1}{c_w^2}\delta_{\mu\nu}$$

$A\phi^+\phi^-$

$$gs_w[q_\mu - k_\mu]$$

$W^-\phi^0\phi^+$

$$g\frac{1}{2}[q_\mu - k_\mu]$$

$W^+\phi^-\phi^0$ $\qquad g\dfrac{1}{2}[q_\mu - k_\mu]$

$Z^0\phi^0 H$ $\qquad ig\dfrac{1}{2c_w}[q_\mu - k_\mu]$

$Z^0\phi^+\phi^-$ $\qquad g\dfrac{2c_w^2 - 1}{2c_w}[q_\mu - k_\mu]$

$W^-\phi^+ H$ $\qquad ig\dfrac{1}{2}[q_\mu - k_\mu]$

$W^+\phi^- H$ $\qquad ig\dfrac{1}{2}[q_\mu - k_\mu]$

$\phi^0\phi^0 H$ $\qquad -g\dfrac{m_h^2}{2M}$

$\phi^-\phi^+ H$ $\qquad -g\dfrac{m_h^2}{2M}$

HHH $\qquad -g\dfrac{3m_h^2}{2M}$

The four-point vertices have no dependence on the momenta. That is typical for a renormalizable theory. Again, vertices becoming non-perturbative for large Higgs mass m_h are marked.

AAW^-W^+

$$-g^2 s_w^2 [2\delta_{\mu\nu}\delta_{\sigma\tau} \\ -\delta_{\mu\sigma}\delta_{\nu\tau} \\ -\delta_{\mu\tau}\delta_{\nu\sigma}]$$

$Z^0 Z^0 W^- W^+$

$$-g^2 c_w^2 [2\delta_{\mu\nu}\delta_{\sigma\tau} \\ -\delta_{\mu\sigma}\delta_{\nu\tau} \\ -\delta_{\mu\tau}\delta_{\nu\sigma}]$$

$AZ^0 W^- W^+$

$$-g^2 c_w s_w [2\delta_{\mu\nu}\delta_{\sigma\tau} \\ -\delta_{\mu\sigma}\delta_{\nu\tau} \\ -\delta_{\mu\tau}\delta_{\nu\sigma}]$$

$W^- W^- W^+ W^+$

$$g^2 [2\delta_{\mu\nu}\delta_{\sigma\tau} \\ -\delta_{\mu\sigma}\delta_{\nu\tau} \\ -\delta_{\mu\tau}\delta_{\nu\sigma}]$$

$AA\phi^-\phi^+$

$$-2g^2 s_w^2 \delta_{\mu\nu}$$

$AZ^0\phi^-\phi^+$

$$-g^2 \frac{s_w(2c_w^2 - 1)}{c_w}\delta_{\mu\nu}$$

$AW^-\phi^+\phi^0$

$$g^2 \frac{s_w}{2}\delta_{\mu\nu}$$

$AW^+\phi^-\phi^0$ $\qquad\qquad g^2\dfrac{s_w}{2}\delta_{\mu\nu}$

$AW^-\phi^+H$ $\qquad\qquad -ig^2\dfrac{s_w}{2}\delta_{\mu\nu}$

$AW^+\phi^-H$ $\qquad\qquad ig^2\dfrac{s_w}{2}\delta_{\mu\nu}$

$Z^0Z^0\phi^0\phi^0$ $\qquad\qquad -g^2\dfrac{1}{2c_w^2}\delta_{\mu\nu}$

$Z^0Z^0\phi^-\phi^+$ $\qquad\qquad -g^2\dfrac{(2c_w^2-1)^2}{2c_w^2}\delta_{\mu\nu}$

Z^0Z^0HH $\qquad\qquad -g^2\dfrac{1}{2c_w^2}\delta_{\mu\nu}$

$Z^0W^-\phi^+\phi^0$ $\qquad\qquad -g^2\dfrac{s_w^2}{2c_w}\delta_{\mu\nu}$

$Z^0W^+\phi^-\phi^0$ $\qquad\qquad -g^2\dfrac{s_w^2}{2c_w}\delta_{\mu\nu}$

$Z^0 W^- \phi^+ H$ $\qquad ig^2 \dfrac{s_w^2}{2c_w} \delta_{\mu\nu}$

$Z^0 W^+ \phi^- H$ $\qquad -ig^2 \dfrac{s_w^2}{2c_w} \delta_{\mu\nu}$

$W^- W^+ \phi^0 \phi^0$ $\qquad -g^2 \dfrac{1}{2} \delta_{\mu\nu}$

$W^- W^+ \phi^- \phi^+$ $\qquad -g^2 \dfrac{1}{2} \delta_{\mu\nu}$

$W^- W^+ H H$ $\qquad -g^2 \dfrac{1}{2} \delta_{\mu\nu}$

$\phi^0 \phi^0 \phi^0 \phi^0$ $\qquad -g^2 \dfrac{3m_h^2}{4M^2}$

$\phi^0 \phi^0 \phi^- \phi^+$ $\qquad -g^2 \dfrac{m_h^2}{4M^2}$

$\phi^- \phi^- \phi^+ \phi^+$ $\qquad -g^2 \dfrac{m_h^2}{2M^2}$

$\phi^0\phi^0 HH$ $\qquad -g^2 \dfrac{m_h^2}{4M^2}$

$\phi^-\phi^+ HH$ $\qquad -g^2 \dfrac{m_h^2}{4M^2}$

$HHHH$ $\qquad -g^2 \dfrac{3m_h^2}{4M^2}$

FP Vertices Involving Vector Bosons, Higgs

The relation between arrows in the FP-lines and charge flow are more complicated than in the above. An outgoing arrow is associated with a conjugate field, such as for example $\bar{X}^-$ (which is not equal to X^+). In that example the charge flow is actually opposite to the arrow. Thus, for a negatively charged FP field the flow of charge is opposite to direction of the arrow, and for a positively charged FP field it is in the direction of the arrow. And for a neutral FP field there is no charge flow.

$\bar{X}^- A X^-$ $\qquad g s_w p_\nu$

$\bar{X}^+ A X^+$ $\qquad -g s_w p_\nu$

$\bar{X}^- Z^0 X^-$ $\qquad g c_w p_\nu$

$\bar{X}^+ Z^0 X^+$

$-g c_w p_\nu$

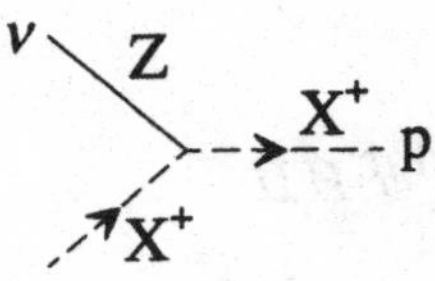

$\bar{X}^- W^- X^0$

$-g c_w p_\nu$

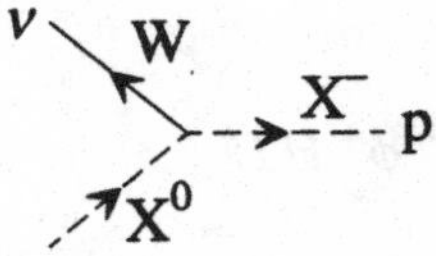

$\bar{X}^- W^- Y$

$-g s_w p_\nu$

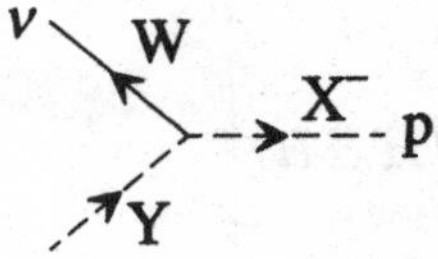

$\bar{X}^0 W^+ X^-$

$-g c_w p_\nu$

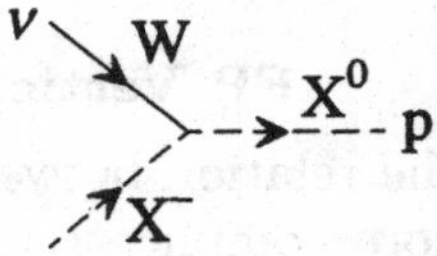

$\bar{Y} W^+ X^-$

$-g s_w p_\nu$

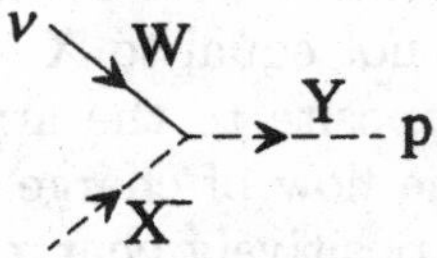

$\bar{X}^0 W^- X^+$

$g c_w p_\nu$

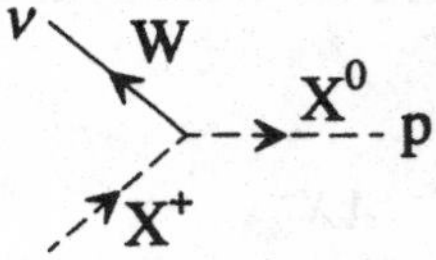

$\bar{X}^+ W^+ X^0$

$g c_w p_\nu$

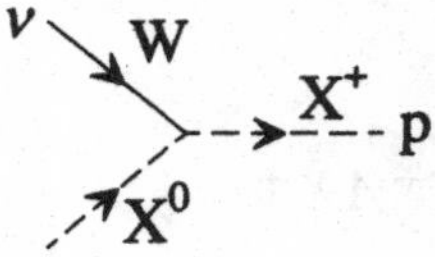

$\bar{Y} W^- X^+$

$g s_w p_\nu$

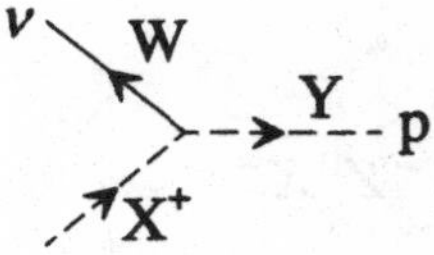

$\bar{X}^+ W^+ Y$

$g s_w p_\nu$

The Higgs-ghost-FP or Higgs-FP vertices have no momentum dependence.

$\bar{X}^- \phi^0 X^-$

$-ig\dfrac{1}{2}M$

$\bar{X}^+ \phi^0 X^+$

$ig\dfrac{1}{2}M$

$\bar{X}^0 \phi^+ X^-$

$ig\dfrac{1}{2c_w}M$

$\bar{X}^+ \phi^+ X^0$

$-ig\dfrac{2c_w^2 - 1}{2c_w}M$

$\bar{X}^+ \phi^+ Y$

$-ig s_w M$

$\bar{X}^- \phi^- X^0$

$ig\dfrac{2c_w^2 - 1}{2c_w}M$

$\bar{X}^-\phi^-Y$ igs_wM

$\bar{X}^0\phi^-X^+$ $-ig\dfrac{1}{2c_w}M$

$\bar{X}^-HX^-$ $-g\dfrac{1}{2}M$

$\bar{X}^0HX^0$ $-g\dfrac{1}{2c_w^2}M$

$\bar{X}^+HX^+$ $-g\dfrac{1}{2}M$

Fermion Vertices

The complications of the Cabibbo–Kobayashi–Maskawa matrix can be dealt with rather easily. Below one will find things such as V_{ud} or the complex conjugate V_{ud}^*. Substituting for example t for u and s for d in the equations gives the expression for top–strange transitions. One uses then correspondingly V_{ts}, equal to $-c_{12}s_{23}-s_{12}c_{23}s_{13}\exp(i\delta)$ according to the CKM matrix representations shown before. At the same time, if there is a mass m_d or m_u, substitute the appropriate mass, m_s or m_t in this example.

The significance of arrow direction with respect to charge flow is precisely as with the FP ghosts, and depends on the charge of the field involved. For up quarks the charge flow is according to the direction of the arrow, for down quarks it is opposite to the direction of the arrow. In the $(\bar{u}dW^+)$ vertex charge from the W distributes: 2/3 flows into the up line, 1/3 into the down line against the arrow direction.

In this part no colour indices are shown, as the couplings are the same for all colours and diagonal in the colours.

$\bar{u}W^+V_{ud}d$ $\qquad ig\dfrac{1}{2\sqrt{2}}\gamma^\mu(1+\gamma^5)V_{ud}$

$\bar{d}W^-V_{ud}^*u$ $\qquad ig\dfrac{1}{2\sqrt{2}}\gamma^\mu(1+\gamma^5)V_{ud}^*$

$\bar{u}Z^0u$ $\qquad ig\dfrac{1}{4c_w}\gamma^\mu\left[1-\dfrac{8}{3}s_w^2+\gamma^5\right]$

$\bar{d}Z^0d$ $\qquad ig\dfrac{1}{4c_w}\gamma^\mu\left[\dfrac{4}{3}s_w^2-1-\gamma^5\right]$

$\bar{u}Au$ $\qquad igs_w\dfrac{2}{3}\gamma^\mu$

$\bar{d}Ad$ $\qquad -igs_w\dfrac{1}{3}\gamma^\mu$

$\bar{u}\phi^+d$ $\qquad ig\dfrac{1}{2\sqrt{2}}\left[\dfrac{m_u-m_d}{M}\right.$ $\left.+\dfrac{m_u+m_d}{M}\gamma^5\right]V_{ud}$

$\bar{d}\phi^- u$ $\qquad ig\dfrac{1}{2\sqrt{2}}\left[-\dfrac{m_u-m_d}{M}\right.$
$\left.+\dfrac{m_u+m_d}{M}\gamma^5\right]V_{ud}^*$

$\bar{u}Hu$ $\qquad -g\dfrac{1}{2}\dfrac{m_u}{M}$

$\bar{d}Hd$ $\qquad -g\dfrac{1}{2}\dfrac{m_d}{M}$

$\bar{u}\phi^0 u$ $\qquad ig\dfrac{1}{2}\dfrac{m_u}{M}\gamma^5$

$\bar{d}\phi^0 d$ $\qquad -ig\dfrac{1}{2}\dfrac{m_d}{M}\gamma^5$

The lepton couplings conclude the weak part.

$\bar{\nu}W^+ e$ $\qquad ig\dfrac{1}{2\sqrt{2}}\gamma^\mu(1+\gamma^5)$

$\bar{e}W^- \nu$ $\qquad ig\dfrac{1}{2\sqrt{2}}\gamma^\mu(1+\gamma^5)$

$\bar{\nu}Z^0 \nu$ $\qquad ig\dfrac{1}{4c_w}\gamma^\mu(1+\gamma^5)$

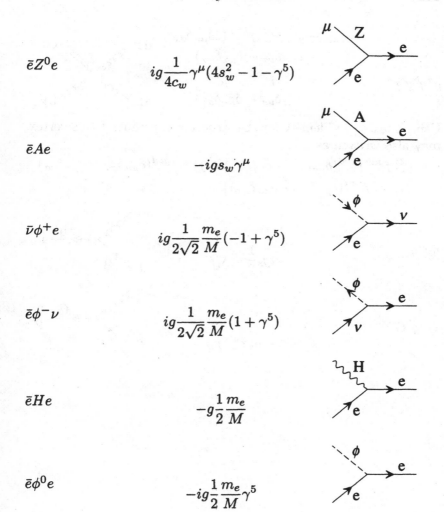

$\bar{e}Z^0e \qquad ig\dfrac{1}{4c_w}\gamma^\mu(4s_w^2-1-\gamma^5)$

$\bar{e}Ae \qquad -igs_w\gamma^\mu$

$\bar{\nu}\phi^+e \qquad ig\dfrac{1}{2\sqrt{2}}\dfrac{m_e}{M}(-1+\gamma^5)$

$\bar{e}\phi^-\nu \qquad ig\dfrac{1}{2\sqrt{2}}\dfrac{m_e}{M}(1+\gamma^5)$

$\bar{e}He \qquad -g\dfrac{1}{2}\dfrac{m_e}{M}$

$\bar{e}\phi^0e \qquad -ig\dfrac{1}{2}\dfrac{m_e}{M}\gamma^5$

Quantum Chromodynamics

$g^ag^bg^c \qquad \begin{aligned} ig_s f^{abc}[&\delta_{\mu\nu}(p_\sigma-q_\sigma)\\ &+\delta_{\nu\sigma}(q_\mu-k_\mu)\\ &+\delta_{\mu\sigma}(k_\nu-p_\nu)] \end{aligned}$

$$g^a g^b g^c g^d$$

$$
\begin{aligned}
- g_s^2 [\, & f^{gac} f^{gbd} \, (2\delta_{\mu\nu}\delta_{\sigma\tau} \\
& - \delta_{\mu\sigma}\delta_{\nu\tau} - \delta_{\mu\tau}\delta_{\nu\sigma}) \\
& + f^{gab} f^{gcd} \, (2\delta_{\mu\sigma}\delta_{\nu\tau} \\
& - \delta_{\mu\tau}\delta_{\nu\sigma} - \delta_{\mu\nu}\delta_{\sigma\tau})\,]
\end{aligned}
$$

Using the Jacobi identity for the structure constants f this vertex may also written as

$$
\begin{aligned}
- g_s^2 [\, & f^{gab} f^{gcd}(\delta_{\mu\sigma}\delta_{\nu\tau} - \delta_{\mu\tau}\delta_{\nu\sigma}) + f^{gac} f^{gbd}(\delta_{\mu\nu}\delta_{\sigma\tau} - \delta_{\mu\tau}\delta_{\nu\sigma}) \\
& + f^{gbc} f^{gad}(\delta_{\mu\nu}\delta_{\sigma\tau} - \delta_{\mu\sigma}\delta_{\nu\tau})\,]
\end{aligned}
$$

$$\bar{q} g^a q$$

$$i g_s \frac{1}{2} \gamma^\mu \lambda_{ij}^a$$

$$\bar{G}^a g^b G^c$$

$$-i g_s f^{abc} p_\mu$$

Appendix F
Metric and Conventions

F.1 General Considerations

Few things touch more raw emotions than the use of metric and other conventions. This is an example of one of Parkinson's laws, stating that trivial issues that everybody can understand are often the most hotly debated ones. It is our intention to discuss and defuse the subject, as it is really a rather trivial issue.

First a piece of history. Einstein himself used different conventions throughout his life. In his work on special relativity he at some point started using Minkowski's metric, or rather, no metric. This is what is now often called the Pauli metric, with imaginary fourth coordinate. In Einstein's book *The Meaning of Relativity*, in the part about special relativity, this usage has persisted unchanged. In discussing gravitation, however, one must of course introduce the metric tensor explicitly, and in 1915 Einstein introduced the $(-1, -1, -1, +1)$ tensor as the zeroth approximation. This same metric has been used by Bjorken and Drell in their book on field theory, and we will refer to this as the bd metric. Einstein never used this metric outside the context of gravitation. It is actually amusing to note the introduction of a third metric, the metric $(-1, -1, -1, -1)$, in his book *The Meaning of Relativity*. Here is his motivation:

"In considering approximations it is often useful, as in the special theory of relativity, to use an imaginary x_4-coordinate, as then the $g_{\mu\nu}$, to first approximation, assume the values

$$\begin{pmatrix} -1 & 0 & 0 & 0 \\ 0 & -1 & 0 & 0 \\ 0 & 0 & -1 & 0 \\ 0 & 0 & 0 & -1 \end{pmatrix}$$

273

These values may be collected in the relation

$$g_{\mu\nu} = -\delta_{\nu\mu}\,."$$

Anyone having worked in gravitation understands this remark, and this becomes even more true if things like vierbeins and spin connections appear on the scene. Anyway, as far as Einstein is concerned we have the choice between three metrics. And rightly so he never really wasted more words on that.

Widely read books in quantum field theory that have promoted the Pauli metric include Sakurai's book, and Källén's books, notably his contribution to the *Handbuch der Physik*. The latter is in German, and that certainly limited its impact in the US. The bd metric owes its popularity not only to the usage by Bjorken and Drell, but also to the (earlier) usage by Bogoliubov and Shirkov, and Feynman. At this time it seems that the bd metric is used by most particle theorists.

In actual fact, when studying the matter more closely, the issue is not so much the use of metric, and the usage of an imaginary fourth coordinate, but rather the question of including factors in defining various objects. In this book the Pauli metric has been used, and one may ask: going through the book, what changes would have to be made to make it according to the bd metric. In fact, we have used more or less randomly x_4 or x_0, it being understood that $x_4 = ix_0$ and the issue of using an imaginary fourth coordinate is essentially a non-issue. The momentum p_μ of a particle was typically something like $(\vec{p}, p_4)$ or $(\vec{p}, ip_0)$, which is, apart from the i, the same as in the bd conventions for the contravariant momentum, i.e., the momentum with upper index, p^μ. In that usage the covariant momentum, a vector with lower index, p_μ, equals $(p_0, -\vec{p})$. In the Pauli metric $p^2 = -m^2$, in the bd metric the dot-product has a minus sign and $p^2 = m^2$.

Lorentz transformations are in the bd metric described by a matrix with one upper and one lower index; the correspondence is as follows. Take a Lorentz transformation as appearing in this book. Remove the i in the last column. Replace the i in the last row by -1. Lift the second index to the upper position. You may want to move the last column and last row to the first position in line with the idea that the corresponding index is 0, not 4.

Then you have the Lorentz transformations according to the bd conventions.

With respect to the representation of the Lorentz transformations in terms of an exponential, with certain generators (the K matrices), the situation is more complicated. Most books do not even write down these things, and in any case there appears not to be any consensus. Some insist on hermitian generators, thereby introducing factors i where there were none in any metric.

Concerning Feynman rules and actual calculations the translation rules are quite simple. In the Feynman rules move all indices upwards, and give all dot-products a minus sign. Replace $\delta_{\mu\nu}$ by $-g^{\mu\nu}$. Replace $\epsilon_{\mu\nu\lambda\sigma}$ by $-i\epsilon^{\mu\nu\lambda\sigma}$. Replace γ^μ by $-i\gamma^\mu$ and γ^5 by $-\gamma^5$. Insert a $-g_{\sigma\tau}$ (thus g with lower indices and a $-$ sign) in every upper index contraction. These rules allow for a line-by-line translation of most if not all calculations in this book.

If all indices are contracted then, apart from the factors $-i$ for γ's and ϵ's and -1 for γ^5, the whole amounts to a minus sign for every contraction (including those implicit in a dot-product) and moving one of the indices down, absorbing the associated $g_{\sigma\tau}$.

Thus all quantities are contravariant. The exception is $\partial_\mu = \partial/\partial x_\mu$ which is covariant as our x is contravariant. Thus ∂_μ translates to ∂_μ, not ∂^μ. And $\partial^2 = \partial_\mu\partial_\mu$ to $-g^{\mu\mu'}\partial_\mu\partial_{\mu'} = -\partial^\mu\partial_\mu$. Now for a lower index contraction we must provide a $-g^{\sigma\tau}$, and nothing at all for a mixed contraction.

Let us explain the origin of these rules for γ matrices and the ϵ tensor. In order to have the bd anti-commutation rule

$$\{\gamma^\mu, \gamma^\nu\} = 2g^{\mu\nu}$$

one must include a factor i in the bd γ matrices compared to ours. The fourth one, γ^4, is the same as the bd γ^0. Thus, $\gamma_p^\mu = -i\gamma_{bd}^\mu$ for $\mu = 1, 2, 3$ and $\gamma_p^4 = \gamma_{bd}^0$. If we had to define a γ^0 it would be such that $\gamma^4 = i\gamma^0$. Thus our γ^0 also equals $-i$ times the bd γ^0. Finally, $\gamma_p^5 = -\gamma_{bd}^5$. We already wrote the indices on the γ matrices in the upper position, so nothing is to be done there. In the various equations for traces one simply replaces γ by $-i\gamma$, $\delta_{\mu\nu}$ by $-g^{\mu\nu}$ (or $-g_{\mu\nu}$ if contracted) and adds $-g...$ as needed. For example, the equation $\gamma^\mu\gamma^\nu\gamma^\mu = -2\gamma^\nu$ becomes

$$i\gamma^\mu\gamma^\nu\gamma^{\mu'}(-g_{\mu\mu'}) = 2i\gamma^\nu \quad \rightarrow \quad \gamma^\mu\gamma^\nu\gamma_\mu = -2\gamma^\nu.$$

Concerning the ϵ tensor there is a little subtlety. The bd ϵ-tensor with upper indices is like ours. Then there is a factor -1 by moving the index 0 to the first position. The result is $\epsilon_p^{1234} = -\epsilon_{bd}^{0123}$. In the bd metric the ϵ tensor with lower indices is minus that with upper indices.

Thus in the Pauli metric $\epsilon_{1234} = 1$, in the Bjorken and Drell metric $\epsilon_{0123} = 1$ and $\epsilon^{0123} = -1$. The ϵ tensor of Itzykson and Zuber differs from that of Bjorken and Drell by a minus sign. Note that in the bd conventions $\gamma_5 = \gamma^5 = i\gamma^0\gamma^1\gamma^2\gamma^3$, and thus $\mathrm{Tr}(\gamma^5\gamma_0\gamma_1\gamma_2\gamma_3) = 4i\epsilon_{0123} = 4i$, while $\mathrm{Tr}(\gamma^5\gamma^0\gamma^1\gamma^2\gamma^3) = 4i\epsilon^{0123} = -4i$.

In the Pauli metric one of the indices of the ϵ tensor is four. That makes it effectively an imaginary quantity, because when contracting it with vectors one of the factors, in the Pauli metric, is purely imaginary. In the translation from Pauli metric to Bjorken and Drell metric that works out altogether to the factor $-i$ mentioned.

Here some examples. If you find in this book the expression $\epsilon_{\mu\nu\lambda\sigma}p_\nu q_\lambda k_\sigma$ then this becomes in the bd metric

$$-i\epsilon^{\mu\nu\lambda\sigma}(-g_{\nu\nu'})(-g_{\lambda\lambda'})(-g_{\sigma\sigma'})p^{\nu'}q^{\lambda'}k^{\sigma'} = i\epsilon^{\mu\nu\lambda\sigma}p_\nu q_\lambda k_\sigma.$$

Suppose in this book you find the expression $\mathrm{Tr}(\gamma^5\gamma^\mu\gamma^\nu\gamma^\lambda\gamma^\sigma)$. In the bd metric this becomes, according to our translation rules:

$$-\mathrm{Tr}(\gamma^5\gamma^\mu\gamma^\nu\gamma^\lambda\gamma^\sigma)$$

There were four factors $-i$ and one factor -1. Now one step further down the trace is worked out, and in this book we find $4\epsilon_{\mu\nu\lambda\sigma}$. According to our rules this translates to the bd metric expression:

$$-4i\epsilon^{\mu\nu\lambda\sigma}.$$

Indeed, that is also what is obtained by doing the trace in the bd metric.

Another example. Suppose in this book you find $\mathrm{Tr}(\gamma^\mu\gamma^\nu)$. This becomes $-\mathrm{Tr}(\gamma^\mu\gamma^\nu)$ in the bd metric. Taking the trace we have $4\delta_{\mu\nu}$, which translates according to our rules to $-4g^{\mu\nu}$ in the bd metric. This is precisely what you find doing the trace the bd way.

Well, that is it as far as metric goes. It really amounts to very little. Then there is the question of other included factors.

Bjorken–Drell have worked out the factors of $(2\pi)^4$ somewhat further than we did. Our rules are:

- a factor $(2\pi)^4$ for every vertex;
- a factor $(2\pi)^{-4}$ for every propagator;
- when squaring the matrix-element: a factor $(2\pi)^{-4}$ from squaring the δ-function for overall conservation of energy–momentum.

Bjorken and Drell prescribe:

- no factors $(2\pi)^4$ for vertex or propagator;
- a factor $(2\pi)^{-4}$ for every loop integral;
- a factor $(2\pi)^4$ when squaring the matrix-element.

The total result is of course the same in both cases, although not for the matrix element itself. Our matrix-element equals that of Bjorken–Drell multiplied by $(2\pi)^4$. After squaring our matrix-element must be provided with a factor $(2\pi)^{-4}$, while Bjorken–Drell prescribe a factor $(2\pi)^4$. A diagram with one vertex and no propagators would, in this book, give a factor $(2\pi)^4$ for the matrix-element; squaring that and remembering the $(2\pi)^{-4}$ from squaring the δ-function for energy–momentum conservation gives $(2\pi)^4$ as result, the same as with the Bjorken–Drell prescription. As another example consider a diagram with one loop, such as the self-energy diagram considered so often. We have no factor $(2\pi)^4$ since there are two propagators and two vertices. But Bjorken–Drell have a factor $(2\pi)^{-4}$ since there is one loop integral. Again, ours equals the Bjorken–Drell expression multiplied by $(2\pi)^4$.

When calculating processes there is thus no difference. But in the unitarity equation $i(T - T^\dagger) = -T^\dagger T$ one must provide a factor $(2\pi)^4$ to the right hand side in the Bjorken–Drell case. A similar remark holds with respect to other factors mentioned below. It is for this reason that we prefer not to take factors from the matrix-element to be compensated after squaring.

Other than that, the propagators and vertices, including the $1/i$ and i that we prescribed separately, are the same. Remember, in the propagators p^2 must be replaced by $-p^2$ when passing to the bd metric. For example, our pion propagator versus that of Bjorken and Drell is:

$$\frac{1}{i}\,\frac{1}{p^2 + M^2 - i\epsilon} = \frac{i}{p^2 - M^2 + i\epsilon}\ \text{(bd)}.$$

To see the profound difference between Pauli and Bjorken–Drell conventions write components explicitly:

$$\frac{1}{i}\frac{1}{\vec{p}^2 - p_0^2 + M^2 - i\epsilon} = \frac{i}{p_0^2 - \vec{p}^2 - M^2 + i\epsilon} \quad (\text{bd}).$$

For a massive vector particle:

$$\frac{1}{i}\frac{\delta_{\mu\nu} + p_\mu p_\nu/M^2}{p^2 + M^2 - i\epsilon} = \frac{-ig^{\mu\nu} + ip^\mu p^\nu/M^2}{p^2 - M^2 + i\epsilon} \quad (\text{bd}).$$

One can use the same expression with the indices down, saving the occurrence of two $-g$ tensors in many cases. A fermion:

$$\frac{1}{i}\frac{-i\gamma p + m}{p^2 + m^2 - i\epsilon} = \frac{i(\not{p} + m)}{p^2 - m^2 + i\epsilon} \quad (\text{bd}),$$

where the Bjorken–Drell notation $\not{p}$ is defined as $\gamma^\mu p_\mu = \gamma^\mu p^\nu g_{\mu\nu}$, which is $-i\gamma^\mu p_\mu$ in our notation. Indeed, according to our rules $-i\gamma^\mu p_\mu$ translates to $(-i)(-i)\gamma^\mu p^\nu (-g_{\mu\nu}) = \not{p}$ in the bd notation. If we also define $\not{p} = \gamma^\mu p_\mu$ then $\not{p}_p = i\not{p}_{bd}$. We prefer for typographical reasons the notation γp instead of $\not{p}$.

Translating vertices to bd conventions one must include the factor i that we prescribed separately for every vertex. For example, including that i our vertex for quantum electrodynamics is $i(-ie)\gamma^\mu = e\gamma^\mu$; according to our rules this translates to $-ie\gamma^\mu$ in the bd scheme.

For scalar and vector particles we have a factor $1/\sqrt{2p_0}$ where Bjorken and Drell have none. They prescribe this factor (squared) later after squaring the matrix-element. They also have no factors V (volume) that we mainly use as a check.

Concerning spinors, there is a factor: the spinors of this book are equal to those of Bjorken and Drell multiplied by $\sqrt{m/p_0}$, with m and p_0 the mass and energy of the fermion. Bjorken and Drell explicitly prescribe the factor m/p_0 after squaring the matrix-element. This is manifest in the expressions for spinors summed over spins; ours differ by a factor m/p_0 from those of Bjorken and Drell. But again, this is corrected by the bd convention that the matrix-element squared must be multiplied by m/p_0. Essentially one may forget about this. Bjorken and Drell denote anti-particle spinors by v.

So much about Feynman rules. There are other conventions concerning inclusion of factors, not necessarily Bjorken and Drell.

There is a factor $\sqrt{2}$ that one may or may not include in the definition of currents; we have mentioned that explicitly where appropriate. Another case is the combination $\sigma^{\mu\nu} = (\gamma^\mu\gamma^\nu - \gamma^\nu\gamma^\mu)/4$ in this book. Everyone else defines this with a factor $1/2$ or $i/2$ rather then the factor $1/4$ that we use. Our usage is such that the σ satisfy commutation rules as required for generators of the Lorentz group. As this σ is otherwise not a frequently used object we have not bothered about this. To avoid misunderstanding, we have usually recalled the definition of σ when using it.

Finally there is the usage to extract a factor $i\pi^2$ in the definition of integrals such as B_0, B_1 etc. as such one-loop integrals invariably have such a factor. We prefer to write everything explicitly if it is not too much trouble. And that is really our attitude: avoid as much as possible all kinds of implicit inclusions.

Incidentally, the functions A and B (and their three and four point extensions C and D) always include as many vertices as propagators. Thus no i's or $-$ signs in Pauli or Bjorken and Drell metric, and they are literally the same in both metrics. At most one could argue a factor $(2\pi)^{-4}$ in the bd metric, which so far seems to have been overlooked as a possibility.

F.2 Translation Examples

The examples below show the application of the translation dictionary specified in the next section.

Lagrangian

This book $\quad -\partial_\nu W_\mu^+ \partial_\nu W_\mu^- - M^2 W_\mu^+ W_\mu^-$

Bjorken–Drell $\quad -\partial_\nu W_\mu^+ \partial^\nu W^{-\mu} + M^2 W_\mu^+ W^{-\mu}$

Remarks: the first term needs two $-g$, the second one.

This book $\quad \partial_\nu Z_\mu^0 W_\mu^+ W_\nu^-$

Bjorken–Drell $\quad -\partial_\nu Z^{0\mu} W_\mu^+ W^{-\nu}$

Remarks: Index on ∂ remains low. Only one $-g$ needed.

This book $\quad -\bar{e}(\gamma\partial + m)e$

Bjorken–Drell $\quad -\bar{e}(-i\slashed{\partial} + m)e$

Remarks: $\slashed{\partial} = \gamma^\alpha \partial_\alpha$.

Feynman rules

Reminder: Feynman rules for the Standard Model in certain publications differ from ours by replacing our $\phi^{\pm}$ by $\mp i\phi^{\pm}$ and ϕ^0 by $-\phi^0$. Designated by "other" below.

Electron photon vertex
> This book $-ie\gamma^{\mu}$
> Bjorken–Drell $-ie\gamma^{\mu}$
> Remarks: vertex i and $-i$ from γ.

$Z^0 W^+ W^-$ vertex
> This book $gc_w[\delta_{\mu\nu}(p_\sigma - q_\sigma) + \delta_{\nu\sigma}(q_\mu - k_\mu) + \ldots]$
> Bjorken–Drell $-igc_w[g^{\mu\nu}(p^\sigma - q^\sigma) + g^{\nu\sigma}(q^\mu - k^\mu) + \ldots]$
> Remarks: vertex i and $\delta \to -g$.

$Z^0 Z^0 W^+ W^-$ vertex
> This book $-g^2 c_w^2 [2\delta_{\mu\nu}\delta_{\sigma\tau} - \delta_{\mu\sigma}\delta_{\nu\tau} - \delta_{\mu\tau}\delta_{\nu\sigma}]$
> Bjorken–Drell $-ig^2 c_w^2 [2g^{\mu\nu}g^{\sigma\tau} - g^{\mu\sigma}g^{\nu\tau} - g^{\mu\tau}g^{\nu\sigma}]$

$\bar{\nu} W^+ e$ vertex
> This book $ig\frac{1}{2\sqrt{2}}\gamma^{\mu}(1 + \gamma^5)$
> Bjorken–Drell $ig\frac{1}{2\sqrt{2}}\gamma^{\mu}(1 - \gamma^5)$

$W^+ \phi^- H$ vertex
> This book $\frac{ig}{2}(q_\mu - k_\mu)$, with q and k from ϕ, H
> Other $\frac{-ig}{2}(q^\mu - k^\mu)$
> Remarks: vertex i and i from ϕ.

$Z W^+ \phi^- \phi^0$ vertex
> This book $\frac{-g^2 s_w^2}{2c_w}\delta_{\mu\nu}$
> Other $\frac{g^2 s^2}{2c_w}g^{\mu\nu}$
> Remarks: vertex i, $\delta \to -g$, i and -1 from ϕ's.

Matrix-elements

The translation amounts to treating γ's etc. as before; in addition take out a factor $(2\pi)^4$ and remove all factors V and $1/\sqrt{2k_0}$.

Pion decay to electron, anti-neutrino
> This book $\frac{(2\pi)^4 i}{\sqrt{2Vk_0}}\frac{1}{V}\{\bar{u}\gamma^{\alpha}(a + b\gamma^5)u\}ik_\alpha$
> Bjorken–Drell $-\{\bar{u}\gamma^{\alpha}(a - b\gamma^5)v\}ik^{\alpha'}g_{\alpha\alpha'}$

Muon decay to $e\,\bar{\nu}_e\,\nu_\mu$

This book $\quad \frac{(2\pi)^4 iG_F}{V^2\sqrt{2}} (\bar{u}_e\gamma^\alpha(1+\gamma^5)u_\nu)(\bar{u}_\nu\gamma^\alpha(1+\gamma^5)u_\mu)$

Bjorken–Drell $\quad \frac{iG_F}{\sqrt{2}} (\bar{u}_e\gamma^\alpha(1-\gamma^5)v_\nu)(\bar{u}_\nu\gamma^{\alpha'}(1-\gamma^5)u_\mu)g_{\alpha\alpha'}$

F.3 Translation Dictionary

To translate from this book to Bjorken–Drell conventions:
- replace $\delta_{\mu\nu}$ by $-g^{\mu\nu}$;
  replace γ^μ by $-i\gamma^\mu$;
  replace γ^5 by $-\gamma^5$;
  replace $\epsilon_{\mu\nu\lambda\sigma}$ by $-i\epsilon^{\mu\nu\lambda\sigma}$ (or $i\epsilon^{\mu\nu\lambda\sigma}$ for Itzykson–Zuber);
- move all lower indices up except on ∂_μ;
- provide all dot-products with a $-$ sign;
  replace γp by $i\not{p}$ and $\gamma\partial$ by $-i\not{\partial}$;
  do nothing to mixed index contractions;
  insert a $-g_{\mu\nu}$ or $-g^{\mu\nu}$ in every remaining upper or lower index contraction (or provide a minus sign for every upper/lower contraction and move one of the indices down/up);
- include the vertex factor i directly in the vertices;
  include the propagator factor $1/i$ directly in the propagators.

At intermediate stages factors are provided at different instances:
- the Bjorken–Drell matrix element equals ours divided by $(2\pi)^4$. That is compensated in the Bjorken–Drell usage by prescribing a factor $(2\pi)^4$ after squaring the matrix-element, as compared to our factor $(2\pi)^{-4}$;
- Bjorken and Drell have no factors associated with external lines for scalars or vector particles, where we have a factor $1/\sqrt{2p_0}$. That is compensated by the Bjorken–Drell prescription of a factor $1/2p_0$ after squaring the matrix-element;
- the Bjorken–Drell spinors equal ours multiplied by $\sqrt{p_0/m}$. The spin-sum expressions are thus different from ours by a factor p_0/m. That is compensated by the Bjorken–Drell prescription of a factor m/p_0 after squaring the matrix-element.

There may be other conventions. For example, the Feynman rules for the Standard Model in certain publications differ from ours by replacing our $\phi^\pm$ by $\mp i\phi^\pm$ and ϕ^0 by $-\phi^0$. That usage appears out of line also with common usage for the σ-model for pions (equivalence theorem).

Index

282

Printed in the United States
By Bookmasters

Printed in the United States
By Bookmasters